Paper Coating Additives

A Project of the Coating Additives Committee of the Coating and Graphic Arts Division
Committee Assignment: 880102.01
Robert J. Kane, Task Group Chairman

TAPPI
Atlanta, Georgia

TAPPI PRESS

The Association assumes no liability or responsibility in connection with the use of this information or data, including, but not limited to, any liability or responsibility under patent, copyright, or trade secret laws. The user is responsible for determining that this document is the most recent edition published.

Within the context of this work, the author(s) may use as examples specific manufacturers of equipment. This does not imply that these manufacturers are the only or best sources of the equipment or that TAPPI endorses them in any way. The presentation of such material by TAPPI should not be construed as an endorsement of or suggestion for any agreed upon course of conduct or concerted action.

TAPPI PRESS
Technology Park/Atlanta
P.O. Box 105113
Atlanta, GA 30348-5113 U.S.A.

ISBN 0-89852-061-4
TP 0102B060

Library of Congress Cataloging-in-Publication Data

Paper coating additives.
p. cm.
Includes bibliographical references and index.
ISBN 0-89852-061-4
1. Paper coatings--Additives.
TS1109.P172 1995
676'.235--dc20 95-7397
CIP

Printed in the United States of America

PREFACE

This TAPPI PRESS book is a revision of the 1978 Paper Coating Additives TAPPI PRESS book edited by Chester G. Landes and Leonard Kroll. With any revision, the past authors and editors contributed significantly to the current publication. Therefore, both the listed authors, reviewers, and past contributors should be credited with this publication.

In the last 20 years, technology has changed in the paper coating industry. The challenge in this publication is to review the past technical information and incorporate current learnings. Therefore, this work represents a bench mark in coating additive technology.

TABLE OF CONTENTS

LIST OF CONTRIBUTORS

Preface

Robert J. Kane, Champion International Corp.

Chapter 1: General Description

Chapter 2: Some Principles of Coating Formulation

Author
Lars G. Andersson, J.M. Huber Corp.

Reviewer
Ken L. Snyder, Mydrin Inc.

Chapter 3: Foam Control

Author
John A. Hill, PPG Industries Inc.

Reviewer
Ted Gammon, Rhone Poulenc Inc.

Chapter 4: Water Retention and Rheology Modifiers

Authors
Teng-Shau Young and
Charles Lee Burdick,
Hercules Inc., Aqualon Division

Reviewer
Anthony Adams, Rohm and Haas Co.

Chapter 5: Colorants

Authors
Terrence L. McComb and
Ralph H. Strickler,
Sandoz Chemicals Corp

Reviewer
Richard Post, RLP Associates Inc.

Chapter 6: Dispersants

Author
Ken L. Snyder, Mydrin Inc.

Reviewer
Lars G. Andersson, J. M. Huber Corp.

Chapter 7: Lubricants

Author
John M. Krasniewski, Jr., Henkel Corp.

Reviewer
Chuck Bunting, Witco Corp.

Chapter 8: Insolubilizers

Authors
Susan F. Boss and William C. Floyd,
Sequa Chemicals, Inc.

Reviewers
Tom Sheerin and Paul Pothier,
Bercen Inc.

Chapter 9: Preservatives

Authors
Stephen D. Goldstein and
Robert W. Cade, Betz PaperChem, Inc.

Reviewers
Tom Surface and Christopher Wiatr,
Nalco Chemical Co.

CHAPTER 1: GENERAL DESCRIPTION *(1)*

Lars G. Andersson

Introduction

Any discussion concerning additives can be classified into the following major areas of interest:

1. Reasons for the existence of additives based upon the need for such products in the operation or the finished product. A philosophy or outlook toward additives is involved.
2. The practical skills and intelligent applications based on experience and observations.
3. Results of applications and testimony of their utilization: some good, some bad.
4. Scientific principles and mechanisms which govern the performance of additives: some we understand; others we do not. This discussion deals primarily with the principles.

In its simplest form, the word additive suggests a wide variety of substances which are combined with other ingredients to modify the chemical or physical properties of various systems. Paper coatings are not first or exclusive in the application or use of small quantities of such special materials which transform an acceptable assembly of ingredients into one of outstanding performance and quality.

The textile industry has scientifically developed finishing treatments for cotton fabric that impart wash/wear properties. This has become important in aiding cotton to maintain its competitive position and actually made possible its survival in the textile market. The essential chemical reaction in these treatments has been shown, with near if not complete certainty, to be a crosslinking of cellulose molecules within the cotton fiber. As in other polymeric systems, the crosslinking increases resiliency and reduces permanent deformation from applied stress. The usual procedure for such treatments is to apply to cotton fabric an aqueous solution of a polyfunctional agent with functional groups that react readily with alcoholic hydroxyl groups. These agents or additives are methylol or hydroxymethyl derivatives of a polyamido compound. Are these principles of cellulose cotton modification applicable to paper with similar results to provide more resiliency under letterpress printing pressure? Is there an analogy between elimination of textile ironing in wash/wear cotton and paper calendering to reduce cockle and wrinkling under stress?

Trace amounts of water in liquid systems have created problems sufficiently serious to warrant the attention of scientists in numerous fields. Chemical water scavengers have evolved as valuable additives for the control of unwanted water in fuel systems. Attempted solutions of problems resulting from moisture in jet fuels, refrigerants, and aerosol propellants have involved such drying techniques as the use of inorganic dehydrating agents, molecular sieves, heat, and absorption. Chemical additives to jet fuel such as 2,2-dimethoxypropane react with water and form by-products which in themselves provide an advantage in preventing gum formation. Water scavengers in dry paper coatings have found useful applications in controlling the electrical properties of coated paper products even under low atmospheric humidity conditions.

Adequate oxidative stability is an extremely important requirement for jet lubricating oils. Oil oxidation is undesirable since it causes viscosity increase, metal corrosion due to acid formation, and excessive oil consumption. Oxidative stability is also necessary in liquid paper coatings and finished coated paper products, especially those with large concentrations of long chain polymers with unsaturated bonds.

Whatever the field of application, these substances, known as additives, perform vital functions and generally represent a small portion of the total system by weight or volume. They are chosen, almost without exception, on a performance basis. Some additives are absolutely necessary to the construction of the formula and the ultimate product design, importing value to the end product. Others are included to solve operating problems, and, quite commonly, the use of excessive amounts will impair the quality of the wet coating or the finished product. Frequently, additives represent that special touch which transforms an ordinary coating into outstanding coating machine performance and final quality.

Rapidly advancing technology in the coated paper and paper board industry makes it extremely difficult to generalize on the many phases of the complex subject of additives from formulation technique to applications. The wide variety of coating machine methods, formulations, finishing equipment, synthetic adhesives, and pigments makes the choice of additives an individual mill selection which best suits the operation, economics, and demands of the grades involved. No mill could possibly use all the additives offered to the market. Many of the additives perform more than one function. Lubricants may act as leveling agents and viscosity modifiers. Defoamers may plasticize and influence flow properties. Viscosity modifiers may also be good dispersants.

Paper coating additives may be defined as substances which are added to the wet coating composition to enhance, optimize, or improve the coating layer properties or to remove operating problems during the mixing, coating, drying, calendering, or finishing processes. All materials in a coating formulation excluding pigments, adhesives, and vehicles can be broadly classified as additives. These materials are used to regulate operating performance and specific attributes of the final coated paper.

Classification of coating additives into clear-cut categories may be based upon any of the following systems:

1. Chemically similar groups.
2. Common end effects which can be achieved.
3. Basic coating component which is modified, i.e., adhesive, pigment, or vehicle.
4. Effect on the wet coating mix or the finished product.
5. Well-understood or poorly understood mechanisms of action.

Each additive, irrespective of mechanism or coating component affected, is ultimately judged or chosen to perform a specific or multiple function. Possible objective classification related to performance results in a number of convenient categories for discussion in this introduction. These remarks are directed toward an understanding of the principles and mechanisms which govern activity in each category. Most of the categories are covered in more detail in subsequent chapters, although not necessarily in the same order, and chapter references are provided as a convenience.

Additive Types and Mechanisms

Dispersants

While there is considerable difference of opinion concerning the need or consistent use for other additives, there is general agreement that dispersants are a necessary part of practically every pigmented formula. Unlike the producers of other coating components, the pigment suppliers have made considerable progress in controlling the additives directed toward their products by incorporating dispersants in the raw material, whether in dry or slurry form. This is especially notable and unique among the kaolin clay suppliers. Similar progress on other pigments represents a challenge for the future. The challenge becomes even greater as we move from pigments to other major coating components.

Numerous chemical compounds, organic and inorganic, are widely used as dispersants. Still, as of today, the polyphosphates are often used for white pigments such as clay, titanium dioxide, calcium carbonate, and talc. Sodium silicates with high silica content used to rank high for this application, but both polyphosphates and sodium silicates are more and more being replaced by organic dispersants such as sodium polyacrylates. These organic dispersants have proven to have good dispersing ability and high resistance to decomposition at high temperature.

Mechanical disaggregation, dispersion, and finally deflocculation are separate unit mechanisms involved in pigment slurry preparation. Most dispersing agents impart electrical repulsive forces to individual pigment particles to counteract normal attractive forces which would cause reaggregation in the finished coating. In a sense, the dispersing agents provide the "lasting power of separation," whereas the mechanical forces are necessary to arrive at the dispersed state.

The repulsive forces of dispersion may be in the form of electrical charges of adsorbed ions furnished by the dissociated dispersing agent or as a film barrier of a protective colloid. In some cases, both mechanisms may be involved.

Some studies reveal that the total repulsive charge is less important for stability than the distribution of charges in the diffuse layer and the interaction between these charged layers which contain dispersant ions. The spatial dimensions of these particles and layers dictate certain theoretical limits of proximity, placing a practical maximum on the solid content necessary for a stable system.

The pigment chemistry, mixing methods, coating, and drying methods influence the selection of kind and quantity of dispersant.

Experiments to evaluate the influence of all components of a formula, including the local water supply, must be conducted to arrive at a successful application.

Additional experiments to determine the possible reversion of the effect upon prolonged storage at various solids and temperatures are also necessary. A typical case is the reaction between calcium ions with polyphosphates which reduces the stability of calcium carbonate dispersions, since the dispersant is slowly combined in this reaction. Much work has been done to provide specialty modified polyphosphates, use stabilizing amines, and apply organic surface active agents for more stability. It has also been clearly proven that phosphate dispersants break down with time and temperature — rapidly at high temperature, e.g. 140–150°F, and more slowly at low temperature such as 90–100°F. These temperatures are often reached in a coating system or clay slurry tank during the summer in the southern United States. Organic dispersants on the other hand are very stable at these temperatures and therefore have found widespread use, either alone or in combination with a polyphosphate dispersant.

Foam Control Agents

Entrained air in a paper coating significantly increases the viscosity of a wet coating system and also causes surface imperfections in the coated paper when the bubbles rupture as the paper dries. Entrained air in the form of isolated bubbles dispersed throughout the fluid, as well as concentration of bubbles at the surface, are troublesome. Entrained air in the body of the coating is more difficult to detect, or to isolate and remove, than that which has accumulated on the surface as a conventional foam.

In complex coating systems, the concentration of surface active agents at the liquid–air boundary tends to lower the interfacial tension and promote the formation of bubbles. This condition alone may not produce a lasting air dispersion because the lower interfacial tension also encourages the coalescence of air into larger, more buoyant bubbles which rise to the coating surface and burst. Therefore, a stable air dispersion also requires a strong film formation at the liquid–air boundary. This assists in explaining our observations that air can be readily dispersed and equally easily removed in a pure liquid slurry system. However, the addition of high molecular weight, film-forming adhesives supplies the second necessary condition for stable air dispersions.

Paper coatings contain all the ingredients to encourage stable air entrainment, or foam. Impurities in natural adhesives, emulsifiers, and dispersants in synthetic latices provide the surface active materials. These important minor components vary in kind and amount with a similar polymer. The differences we observe in forming tendency of a coating system are caused by these "additives," not the polymer itself. The adhesive producer uses "additive" to solve his operating problems of modern synthesis, which in turn creates foam problems for the coater, forcing the use of antiadditive additives.

Is it reasonable to direct our research toward the elimination of the true causes, rather than to continue our treatment of the symptoms?

It appears more desirable to avoid the problem of air entrainment by handling the coating systems with a minimum of air admixture during coating preparation, distribution, and application at the coating machine. Although much progress has been made and more is certain to come concerning mechanical improvements, chemical agents are used to eliminate entrained air. Ready classification exists as "antifoams" to prevent stable air dispersions and "defoamers" to combat air dispersions which form despite our best preventative efforts. Some compounds are effective as dual purpose agents. The coating formulator or process operator must determine which material and how much is required to fit a specific set of circumstances. The signals which indicate a need for air control additives are much more obvious than those which reveal that a potential troublesome excess has been used.

Fundamentally, it is believed that antifoam agents discourage the formation of stable surface films around the bubbles. Much remains to be learned regarding the mechanics of this action. Defoamers basically lower the interfacial tension to promote the coalescence of small bubbles into larger bubbles which rise to the surface and burst.

The challenge for natural and synthetic adhesive suppliers to incorporate foam-control agents, as the pigment suppliers have incorporated dispersing control agents, is readily apparent. Low-foaming adhesives alone are not enough; nonfoaming complete systems are required.

Preservatives

Coatings containing adhesives, emulsifiers, stabilizers, and plasticizers supply basic elements such as nitrogen and carbon which are necessary to support life and reproduction of living microorganisms. Recirculation, aeration, and appropriate incubation temperatures provide an ideal environment to encourage proliferation of bacteria and fungi. Microorganism attack on the adhesive can change the viscosity, flow properties, and bonding strength. Organism excretion can cause pH changes, dirt specks, discoloration, and malodor. Preventative action is required to combat this problem because, once started, the effect of microorganism damage is irreversible. Good housekeeping, periodic sterilization, and good equipment design aid in minimizing this problem.

Certain chemicals act as toxicants to the microorganisms by rupturing the organism cell wall, penetrating the cell walls and interfering with oxidation of carbohydrates and other mechanisms which stop the metabolic process.

Chemical classes of microbial toxicants include phenolics, thiones, thiazolins, organic sulfurs, organo halogens, amines, quaternary ammonium salts and organo-metallics. Each preservative problem requires a microbiological study to isolate and identify the species of organism and to discover the proper toxicant, method of addition, and the quantity required to combat the colony.

Insolubilizers

Wet calendering, offset printing, outdoor exposure, and water contact with labels represent some of the many reasons why coatings must be resistant to moisture. The rate at which resistance is developed and the degree required cover a wide range. Extremes are represented by the resistance necessary for offset printing compared to boxboard or tag coatings which must be waterproof for outdoor exposure. Paper to be wet calendered on a board machine, in a continuous process, requires immediate insolubilization, whereas paper stored prior to conversion may permit a slower development.

Chemical reaction between the coating adhesive and the "additive" is the usual mechanism to obtain water resistance. Methods of application vary but include addition to the coating before application or incorporation in a prime or sub coating, as a wash coat, as a vapor treatment in the dryer, and as a spray application. The logical solution of method depends on the adhesive, additive used, coating method, and end-use demands. Ideally, addition to the wet coating system is preferred, provided undesirable changes in rheology can be avoided.

Starch, protein, and synthetic adhesives vary in their inherent ability to resist water damage. Starch and protein coatings can be insolubilized by blending with resins or latices which are water resistant. The water-sensitive chemical groups of the adhesive can be reacted with a resin which blocks the tendency to go into solution. Another mechanism is to use crosslinking chemicals that cause the adhesive to polymerize and reduce the tendency to redisperse in water.

Starch insolubilization has been accomplished by reactions with formaldehyde which is acid catalyzed to yield water-insoluble compounds. Condensation products of formaldehyde with phenols, melamine, and urea develop suitable wet-rub resistance with starch upon aging. It is postulated that the aldehyde and the hydroxyl groups of the starch form a complex insoluble compound. The crosslinking of starch molecules proceeds until the starch loses its ability to swell in water. Recently, however, formaldehyde-containing or formaldehyde-emitting compounds have increasingly come under attack from OSHA or from certain environmental groups and will be phased out in the near future. Therefore formaldehyde free or non-emitting products like AZC (ammonium zirconium carbonate) or cyclic amide condensates have gained widespread acceptance.

The most important chemical property of protein available to the paper coater is the carboxyl acid reactive and the amino basic reactive nature of the molecules. This reactivity makes it possible to form soluble salts with alkalies and form solutions. The molecule is also reactive with organic compounds such as formaldehyde or aldehyde donors. Aldehyde reacts by replacement of an active hydrogen atom of the amino acid group in the protein molecule.

The basic reaction involved between the aldehyde and the hydroxide group in polyvinyl alcohol is also responsible for insolubilization. The aldehyde group may be supplied by any donors, the selection depending upon operating conditions, preparation, and economic factors.

Antistatic Agents

The accumulation of static electricity in paper results in cohesive forces which cause the web to adhere to nonmetal rolls and make sheet separation difficult in the paper and printing plant. Static electricity can be eliminated by the addition of chemicals which make paper conductive, even in the absence of excessive moisture. Antistatic agents can be used to ease operating problems on conventional papers and also represent a method to obtain conductivity on electrophotographic functional papers. Generally speaking, there are three major mechanisms involved in antistatic treatment. Some materials such as glycerides and glycols simply increase the hygroscopicity of the coating, thereby assuring conductivity of the coating to speed the dissipation of the electrical charge. Chemicals also tend to induce an opposite charge to the charge which normally builds up on a material itself. In some instances, the charge may be positive and, in other cases, negative. Which it will be depends upon the orientation of the molecules in the outer layer of the surface active agent.

Ordinary untreated coated paper has a very low electrical conductivity, especially when it is dry. At 12% relative humidity, the resistance of untreated paper is more than 37,000 megohms. Inorganic salts such as potassium chloride or other conductive materials can be added to the paper coating to increase conductivity. Salts, although used, do not provide satisfactorily high conductivity at low humidities even with the aid of humectants. Organic quaternary ammonium compounds dependably lower electrical resistance of coated paper.

Colorants

A large part of production of coated papers is concerned with colored book papers, labels, bristols, covers, and identification tags. The choice of colorants depends upon the requirements of the finished product. Properties such as bleed, tinctorial strength, brilliance, reactivity, fade resistance, and cost must be considered by the formulator in choosing a color combination. In some instance, the entire total dry pigment content may be colorant. In others, only a small percentage may be required. Pigments or water-insoluble materials used for coloring water-based mineral coatings are classified as inorganic and organic compounds. Natural and synthetic products occur in each class. Water-soluble dyestuffs, natural and synthetic, are also used where soluble compounds are permissible.

Most inorganic pigments are primarily iron oxides and hydroxides. Yellow pigments are derived from ocher, raw sienna, and natural oxide. Browns are obtained from raw or burnt umber. Synthetic inorganic pigments include lead chromates which range in color from yellow through orange.

Soluble organic dyes are sometimes combined with inorganic white pigments which constitute a class of colorants known as color lakes or pulp colors. These are sometimes produced by precipitating dyes upon an insoluble base by using a metallic salt or other reagent. These color lakes are available in a wide range of shades and, unlike the soluble dyes from which they are derived, they become insoluble in water and have considerable light fastness.

Aluminum, copper, and bronze metallic powders are also used as coating pigments to obtain special color or functional effects in some coated papers.

Optical Modifiers

The higher whiteness of many current grades of paper improves printed legibility and gives crisper, sharper illustrations. Chemical compounds composed of stilbene-cyanuric derivatives are substantive to coating pigments and act as whitening additives. These compounds function by absorbing ultraviolet light and re-emitting it as visible, blue–white light. The blue–white light neutralizes yellow casts, and this emission increases the overall apparent reflectance of the paper, thus a whiter white. These additives supplement and increase the whiteness and brightness obtained by the normal methods of pigmentation.

Ultraviolet radiation absorbers are used to stabilize the organic components of coatings against the harmful effects of radiation. Shorter wavelength, energetic, unfiltered solar energy which is incident upon an opaque coating will be reflected or absorbed. If it is reflected, the energy is diverted and no temperature rise will result. The additives provide highly absorbing molecules which have a very low tendency for undergoing photochemical reaction. Salicylates, benzophenones, and benzotriazoles are commercial compounds which exhibit these properties.

The ability of a compound to resist photochemical degradation is determined by the susceptibility of the chemical bond to sever upon irradiation and the method by which the materials dispel the absorbed energy (for example, by infrared radiation).

The general requirements for an effective ultraviolet absorber include: (a) photochemical stability to radiation while in the excited state, (b) compatibility with the polymeric material in which it is to be used, (c) little or no evaporation from this material, and (d) strong absorption in the wavelength region for which the polymer will normally degrade.

Gloss agents are materials which increase the angular reflectance of light on the coated surface. Coatings are sometimes formulated with an excessive amount of adhesive to ensure a high pick which makes it more difficult to obtain gloss. Some gloss additives function by lowering the viscosity and causing increased vehicle penetration, leaving more unbound pigment on the surface of the coating which promotes gloss. Plasticizers also contribute toward improved gloss by enabling the adhesive film to become more flexible and thereby yield to pressure and accept a finish more readily. Plastic pigments (usually polystyrene) readily produce gloss during finishing, but are not true additives and will be discussed elsewhere.

Flow Modifiers and Water Retention Aids

To ask how a coating will respond to forces is surely a question of fundamental importance. We want to know how a coating will flow under a mixing load, how it will flow through pumps and pipes, and how it will react to forces of various coating methods. Before we can hope to provide answers to these questions, it is necessary to characterize the mechanical properties of the coatings with which we are concerned. During the last 20 years, a very significant advance has been made in our theoretical understanding of how these coatings behave. Predicting how coatings will behave in response to forces lies in the domain of rheology — literally the study of flow.

Chemical flow modifiers are added to wet coatings to control, lower, and stabilize the viscosity of adhesives and to regulate the flow of the finished coating mixture. These additives provide additional latitude to the formulator and permit a wider selection of raw materials to meet certain end requirements without impairing the flow properties in actual operation. Some materials increase and others decrease viscosity. An occasional product such as polyvinylpyrrolidone can do both, depending upon the conditions of usage and the molecular weight.

Agents which modify flow through the mechanism of coating and vehicle holdout, evening and leveling, and viscosity adjustment can all be conveniently grouped as Flow Modifiers.

The need for adjustment of the rheological properties of a coating varies with the coating application method. In general, coating rheology in a high-speed blade coater, e.g., is more critical than in an air knife or pigmented size press, primarily because of the high solids and speed in the former. The recent drive toward higher and higher blade coating solids has also made many coatings run on the border of dilatancy, making the high-shear rheology most important. LWC paper, usually open and absorbent and produced at high speed in a short dwell coater, will require a coating with low Brookfield and Hercules viscosity, as well as good water retention for trouble-free production.

Coating Bond Modifiers

Improvement in dry and wet coating bond without the addition of surplus adhesive has long been an objective of paper coaters. Dry pick agents take the form of catalysts which promote crosslinking or polymerization of natural and synthetic adhesives, holdout agents which prevent loss of adhesive into the basestock, and plasticizers which prevent adhesive embrittlement and deposit organic films on the surface which bond minerals in the uppermost layer. Adequate understanding does not exist as to whether "non-adhesive" additives improve the actual cohesive forces of the coating particles or whether they decrease the adhesive forces between the ink film and the coated surface.

Wet pick agents function through the mechanism of rendering the major adhesive film less water sensitive or by providing an insoluble film supplement to protect the soluble adhesive. Insoluble synthetic latex and resin films can, to a certain extent, be classified as such additives.

Coating Surface Performance Modifiers

A need sometimes exists to modify the coated surface to regulate gloss ink holdout, grease and oil resistance, scuff and abrasion resistance, and rewetting properties. The mechanism in all such applications is surface activity involving a strong tendency of a solute to concentrate at an interface. Surface active molecules have two distinct portions: One with strong affinity for the solvent, and the other rejected by the solvent. When the repulsion forces are strong, the solute molecules tend to concentrate at the coating surface-air surface interface. Ordinary soaps behave this way when dissolved in water, the polar group being attracted by the water and the hydrocarbon chain being rejected. This causes a concentration of the soap at the water–air and coating–air interface during drying.

Numerous compounds which behave according to the above description have been synthesized and offered to the coating industry. These surface active compounds are classified as follows:

1. Anionic compounds in solution with the long chain negatively charged
2. Cationic compounds in solution with the long chain positively charged
3. Nonionic compounds which do not ionize in solution
4. Amphoteric compounds in solution with the long chain either positive or negative, depending on the pH of the solution.

Common sodium and potassium soaps, ammonia and amine soaps, alkyl aryl sulfonates, and sulfated vegetable oils are examples of chemical classes which perform a variety of end functions.

Lubricants

During the coating operation, lubrication of the wet coating can improve the flow properties. There is a definite degree of overlapping in the action of flow modifiers and lubricants, particularly as they affect the wet coating mix. Lubrication of the dry coating during calendering, cutting, trimming, and folding can affect smoothness, finish, printability, resistance to cracking, and degree of dusting. Paper coating lubrication mechanisms can be divided into two categories. The first involves hydrodynamic lubrication and occurs when moving pigment surfaces are completely separated by a water layer. The second is boundary lubrication. Here there is paper surface and process machinery contact where absorbed films of lubricants reduce friction between the respective surfaces.

Additives which lubricate in the fluid coating phase may improve flow properties and alter the total rheology. Surface lubrication influences the finished coated paper properties. The same materials which were cited as coating surface modifiers are also included as lubricants. Calcium stearate is still a very common lubricant today. It is a common belief that the calcium stearate performs a lubricating function under the blade in a blade coater as well as in the super calender nip, where it reduces the dusting tendency. Calcium stearate can produce an unwanted viscosity increase in high starch coatings, and is therefore often replaced by, e.g., polyethylene emulsions or modified glycerides in these instances.

General Considerations

Organization of the preceding discussion relative to coating component modified and point of effectiveness is shown in **Table 1.1.**

Table 1.1. Performance Modifiers and Their Effect

Component Modified	Point of Effectiveness	
	Process	Finished Product
Pigment	Dispersant (1)	Bond Modifiers (8) Lubricants (10)
Adhesive	Air Entrainment Agents (2) Flow Modifiers (7) Insolubilizers (4) Preservatives (3)	
Total System		Antistatic (5) Optical modifiers (6) Surface-Active Agents (9)

Consideration is given to primary or dominant influence only. Numerous secondary or tertiary effects may also be possible.

It appears that our understanding of the principles or mechanisms and our ability to control the performance decreases from category one to ten, shown in parentheses in Table 1.1. Actual ranking, relative to scientific understanding, may vary among different observers. However, it appears that the additives directed toward the process are better understood and regulated than those directed toward the finished product.

As we direct our effort with additives toward process improvement to broaden the scope of what we *can do,* we must be constantly alert to what the *customer wants.* Additives which solve a particular operating problem must be thoroughly evaluated by asking some of the following questions:

1. Has the additive changed the ink holdout or absorption properties?
2. Do these compounds create ink repellency with age?
3. Has there been an effect on the adhesive to cause loss in pick resistance with time?
4. Has the ink set time, drying time, and ink-to-paper bond been adversely influenced?
5. Do some additives cause fuming stains during ink drying in a stack of printed sheets?
6. Is the long-term color stability of the finished product affected?

Suppliers of additives have contributed much to providing materials that solve coating problems. More remains to be done to investigate the effects of these products on the printing performance of the coated sheet.

There is still much "black art" associated with the use of coating additives. It is hoped that this monograph will stimulate the type of research required to place their use on a more scientific basis.

Literature Cited

1. Bush, T. W., "Paper coating additives," *Tappi* 49(6): 154A–158A(1966).

Acknowledgment

The author specifically wants to recognize Thomas W. Bush whose material in the previous monograph has been used to a very great extent.

CHAPTER 2: SOME PRINCIPLES OF COATING FORMULATION

Lars G. Andersson

Introduction

A coating color is a combination of ingredients each of which is selected to fulfill a specific function. However, as these ingredients are combined in a coating color, they will interact physically and chemically, and therefore each has to be selected keeping in mind its effect on all others. The successful formulator is the one who knows how to balance the ingredients and how to select the ones that interrelate most efficiently. In a good coating color, the total effect is more than just the sum of the individual ingredients. The total is greater than the sum of its parts.

The basic considerations are the requirements of the end product. Generally these involve both the appearance of the product as well as its performance, such as printability. Just as important are the constraints imposed by the manufacturing operations including rawstock characteristics, viscosity limitation, and calendering limitations. With increasing blade coater speeds, high-speed runnability is becoming increasingly important. Finally there are the economics, since a coating is of little value unless it results in a product which can be sold at a profit.

In view of the large number of variables involved, especially the multitude of ingredients from which a choice has to be made, it is never practical to reach the absolute optimum before commercializing a formulation. If perfection is desired before introducing a new coated product, an inordinate amount of experimentation will be required and much potential revenue from an adequate product will be lost while the search for perfection goes on. Possibly a competitor will steal the market. Thus an important part of formulation is the judgment of the point where the product is ready for the market.

However, experimentation and exploration should never cease. Evolutionary Operations (EVOP) is one technique used to optimize the product. Surveillance of market performance and customer comments will serve as alerts when a product is in need of upgrading. New materials are continually being introduced and should be classified in the laboratory to build up a background of data for the day when a specific request for change or improvement is made.

One danger is that coating formulations can become excessively complex. A new material is found which eliminates an operational problem, and it becomes part of the formulation. Does it make obsolete a material which was added some time ago for the same purpose but was not quite as effective? Often the history of a formulation is lost; new ingredients are added, but old ones are rarely taken out. A periodic review of all formulations is therefore advisable, not only to prevent their becoming too complex, but also to pinpoint any shortcomings which may serve as objectives of new development projects.

Process Control

Process control must be practiced with coating colors as with any other manufacturing process. A good control system consists of three basic steps. First the formulations must be clearly defined including order and rate of addition. Next comes the monitoring of these procedures with recorders and operating audits. Finally it is good practice to routinely test the raw materials as well as the finished coating color, selecting tests which represent significant characteristics. No coating operation should run without at least monitoring the rheology, solids, pH, and ash content of the coating colors, as well as basic testing of incoming raw materials.

In view of the interdependence of the criteria which a coating color must meet, it is difficult to describe a procedure for formulating a new coating which will be satisfactory for all occasions. However, there is a general approach which can be applied to most cases.

End Product Requirements

First consider the end product requirements, both for performance and appearance. The pigment and binders as well as their proportions are then selected to meet these requirements. This does not mean that the types and amounts can be selected by reference to the literature or to experience. Rather the formulator will select the most likely candidates and design an experimental program around them.

This can become an extensive project due to the large variety of ingredients which have become available in recent years, a veritable supply explosion. Whereas a few years ago there was a simple choice between three basic latexes for print coatings - acrylic, acetate, and styrene/butadiene, the choice has been broadened by the introduction of many copolymers. More important, many of these types of latex are now available in an alkali-swellable or alkali-soluble form, a characteristic which has improved water-holding ability to the point where it is often feasible to operate with a coating based on synthetic binders alone. Some of these alkali-sensitive binders are designed to be used as sole binders, others to be used in conjunction with conventional latexes, serving as a water-holding agent which at the same time has appreciable binding efficiency.

Economics used to dictate the choice of binders in many instances because significant price differentials existed between various latexes. However, these differences have been reduced to the point where cost considerations are less critical. The latexes are now priced competitively with each other, and with isolated soy protein. Starch is the only major binder which is significantly cheaper.

There are so many varieties of raw materials and end-use requirements that it becomes risky to generalize. For example, in the past it was considered essential that the surface of an enamel coating be alkaline. This was achieved by always including at least a token amount of calcium carbonate pigment in the formulation. Many successful formulations in use today do not include calcium carbonate, and they do not present an ink drying problem. The improved ink receptivity of the newer latexes is considered to be the reason why this change in surface requirements has taken place.

However, with the recent onset of alkaline papermaking and the rapid growth of on-site PCC plants, things are changing again. More and more calcium carbonate is being used in the coating — still primarily the wet ground type — but with special coating grades (rhombohedral and rosette shaped) of PCC gradually coming into use.

General Functions of Additives

Coating color ingredients other than binders, pigments, and water are generally considered to be additives. Their nature and function are described in detail in other parts of this monograph. While the pigments and binders are selected to yield a basic coating color which will achieve the desired end-product characteristics, the general function of additives is to make the basic coating color compatible with the rawstock and the coating equipment, but also provide control of foaming, lubricity, crosslinking, rheology, and water retention.

Additives selection must be made keeping in mind all other aspects of the coating color. A case in point is the pigment dispersant. This is particularly important due to the fact that most coating clays are "predispersed," i.e., have dispersants attached to their surface which are residuals from the process used in their manufacture. Thus pigments of substantially equivalent particle shape and size distribution may be quite different in their dispersion characteristics. In a typical coating color, one clay may need no additional dispersant added, one works best with an inorganic dispersant, and a third need an organic dispersant. It is even possible to find a pigment which appears incompatible because a reaction of its dispersant with the latex system causes pigment shock. While this is an unusual though real situation, should such a pigment–binder combination be the initial one selected in developing a new coating, a good idea may be discarded unless the formulator is aware of this possibility. The motto of formulation is "be alert to interactions." Compatibility may be achieved by using another combination of dispersants.

Both latexes and pigment slips are disperse systems depending on surface charges for their stability. When two such dispersions are mixed, it is difficult to predict what will happen — except, of course, in the case of dispersions with opposite electrical charge where coagulation is assured. Fortunately latexes and pigment slips are generally anionic.

In developing a coating formula, it is common practice to prepare a series of pigment slips, evaluating one or more dispersants to find the type and amount which gives minimum viscosity. To this slip, the various binders are then added in different ratios, studying the viscosity of the final color and possibly the properties of coatings made by laboratory drawdowns. This procedure overlooks an important interaction, namely the possibility that the pigment may "rob" the latex of its dispersant or vice versa. Hence, the evaluation of dispersants must not stop once the optimum has been found for the pigment slip. The proper type and quantity must be verified for the final coating color. The following checklist has been found useful for keeping track of factors to consider while formulating:

1. End Product Requirements

Printing process (sheet offset, web offset, letterpress, gravure, etc.)

Other treatment (varnish, bronze emboss, die cut, etc.)

Physical (strength, stiffness, etc.)

Optical (opacity, brightness, shade, etc.)

2. Rawstock Constraints

Furnish selection

Internal sizing

Surface sizing

Formation

3. Equipment Constraints

Rheology

Water holding

Drying rate

4. Cost Constraints

Raw materials

Coater speed limitations imposed by formulation

Material-handling costs

A more detailed discussion of coating formulation principles has been provided by Nadelman and Baldauf *(1)*. Additional information on current formulating practices is given by Gunning *(2)* and Loomer *(3)*.

Literature Cited

1. Nadelman, A. H., and Baldauf, G. H., *Coating Formulations-Principles and Practices,* Lockwood Trade Journal Co., New York, 1966.
2. Gunning, J. R., "Coating," in *Pulp and Paper Manufacture,* 2nd edn., Vol. II (R. G. Macdonald and J. N. Franklin, Eds.), McGraw-Hill, New York, 1969, Chap. 9, pp. 457–532.
3. Loomer, J. T., "Coated Paper," in *Handbook of Pulp and Paper Technology,* 2nd edn. (K. W. Britt, Ed.), Van Nostrand Reinhold, New York, 1970, Chap. 7-1, pp. 517–528.

Acknowledgment

The author specifically wants to recognize Gunther H. Baldauf whose material in the previous monograph has been used to a very great extent.

CHAPTER 3: FOAM CONTROL AGENTS

John A. Hill

Introduction

The past decade has seen major changes in the design of paper coating makedown and supply systems along with similar changes in coating applicators. The design changes that have occurred have resulted in relatively closed systems; e.g., continuous makedown, in-line pressure filters, and the short-dwell applicator. From the design aspect, the modern makedown and coating supply system does not have as many points where air can be entrained into the coating color, thereby contributing to the problem of foam control.

Offsetting this development has been the constant increase in operating speed within the same period. One of the greatest contributors to air entrainment is the paper web itself continuously carrying air into the coating color at the point of application. Any increase in coater speed must be accompanied with a similar potential for increased air entrainment in the coating color.

From an idealistic standpoint, it would seem that the problem of controlling air entrainment and foam has diminished as a result of improved design and, to some extent, that is correct. Practically speaking, however, it is probable that improved design, although known, is not always practiced; and the need for foam control agents is as great as ever. It is likely that overall control of foam has improved because air entrainment is more likely to occur within the applicator/supply loop, and not as a result of coating makedown. It is also probable that coating formulations are less prone to foam and entrain air due to improved raw materials.

This may be somewhat academic because not all coating facilities are new or have benefited from improved design. It is probable that, at this point in time, air entrainment in coating can be prevented to a much greater extent, provided good equipment design is used and coating colors are formulated to best utilize that design.

Understanding Air Entrainment and Foam Control

Busch *(1)* provides a concise and easily understood explanation of the mechanism that results with entrained air in a coating.

"Entrained air in paper coating significantly increases the viscosity of a wet coating system and also causes surface imperfections in the coated paper when the bubbles rupture as the paper dries. Entrained air in the form of isolated bubbles dispersed throughout the fluid, as well as concentration of the bubbles at the surface, are troublesome. Entrained air in the body of the coating is more difficult to detect, or to isolate and remove, than that which is accumulated on the surface as a conventional foam.

"In complex coating systems, the concentration of surface active agents at the liquid–air boundary tends to lower the interfacial tension and promote the formation of bubbles. This condition may not produce a lasting air dispersion because the lower interfacial tension also encourages the coalescence of air into larger, more buoyant bubbles which rise to the coating surface and burst. Therefore, a stable air dispersion also requires a strong film formation at the liquid–air boundary. This assists in explaining our observations that air can be readily dispersed and equally easily removed in a pure liquid slurry system. However, the addition of high molecular weight film-forming adhesives supplies the second necessary condition for stable air dispersions.

"Paper coatings contain all the ingredients to encourage stable air entrainment, or foam. Impurities in natural adhesives, emulsifiers, and dispersants in synthetic latices provide the surface active materials. These important minor components vary in kind and amount with a similar polymer. The differences we observe in foaming tendency of a coating system are caused by these 'additives,' not the polymer itself."

Adhesives used in paper coatings are major contributors to air entrainment and a comparison of the foaming tendency of one coating to another, as Busch has suggested, will make obvious that certain adhesives are more likely to foam than others. But the problem is oversimplified if the adhesive is considered to be the only source. If, in changing adhesive, a mill experiences a significant increase in foam, it is likely due to that change and a corresponding foam control agent change can alleviate the foam problem. But if the mill has routinely had a foam problem or it is a startup situation, it is not at all certain that the source of the problem is the adhesive. Air entrainment may begin at the first step in coating makedown, long before the addition of adhesive, when only the pigments and dispersants are present in the makedown vessel. If allowed to continue until after the natural adhesive is added and the protective colloid effect introduced, air entrained in that initial step may persist throughout the coating process. When air entrainment is allowed to happen in the presence of moderate or high-shear mixing and when protective colloids are introduced, it is quite likely to be impossible to remove that air in the normal time allotted in a coating mill.

Air entrainment can occur at any step in the coating process due to poor equipment or system design, lack of maintenance, and operational practices. Air entrainment to some degree is expected from those sources and can usually be controlled with the use of the proper foam control agent. An unidentified or intermittent air entrainment source is difficult to control. For that reason, the first step in a foam control evaluation is a careful visual survey of the coating makedown, storage, and application systems. Usual trouble spots are worn pumps that show leakage and are likely to "suck air," service tanks with more than adequate mixing and/or turbulence, or varying operating levels within the service tank. Problems of this sort, if out of control, may make it impossible to establish adequate control of air entrainment with use of a foam control agent without creating the typical problems associated with overuse, i.e., "fisheyes," "defoamer streaks," deposits, etc.

Two additional factors in air entrainment control relate to pH and microorganisms. Generally, foam control agents function more effectively in coatings that are alkaline, preferably pH 8–9.5. It is not unusual to see operations that are run at a lower pH. The pH is measured at the point of coating makedown. When the coating is applied, the pH has dropped to near, or below, neutral. At neutral pH and at a point distant to the usual biocide addition point, microorganisms can thrive, further lowering pH. As a result of the degradation process, emitted gases behave the same as entrained air. With this condition, foam control agents will not function effectively. This condition may occur intermittently because it will be alleviated with a coating system boilout and recur in time.

Methods of Control of Air Entrainment and Foam

Usually, the most successful control of entrained air and foam is neat addition of foam control agent in the initial stage of coating makedown. As stated in "Understanding Air Entrainment and Foam Control," pigment and dispersant alone will entrain air, and it is important to prevent air entrainment throughout the coating process to obtain optimal results. Early addition of foam control agent allows the use of the most hydrophobic foam control agent the coating will tolerate, relying on the shear of makedown and time to thoroughly disperse the chemical before application. Sufficient time and shear is necessary to prevent undesirable side effects such as voids ("fisheyes" and "defoamer streaks," etc.). This is not to say that certain systems will not need additional foam control agent added later in the process, but the best chance of having good control of air entrainment and foam is obtained by early addition of a highly active foam control agent. Selection of the proper foam control agent for application in this manner may require extra effort, but the results usually are worth the effort.

Foam control agents are also added continuously into the coater recirculation system, either into the return line at the coater, or into the service tank. Results of application in this manner can be excellent, but also potentially more difficult in selecting the correct agent. Compared with a makedown agent, the recirculation agent must disperse much more readily to avoid creating coating defects and generally have a "lighter touch." A shortcoming of this method is that it may be difficult to manage in a large mill because control of the coating chemistry, to some extent, is taken away from the group that understands it best, the coating prep department. When foam control agent is added at the coater, the agent required must have quick "knockdown" — "hold-down" is of secondary importance. The agent may be added neat or as a 5–10% emulsion. Neat addition, particularly in recirculation, requires a very mild agent that disperses readily to avoid creating coating defects. A foam control agent to be applied as an emulsion requires outstanding emulsion qualities. Emulsion preparation is elementary, but requires strict adherence to a set procedure. Usually, application in this manner requires a quantity of emulsion prepared each shift and the emulsion continuously stirred to maintain uniformity. The specifics of this application depend upon the product to be used. Emulsion qualities vary between products and must be assessed by the chemist/supplier involved.

Mechanical devices to remove entrained air or provide high-shear makedown without entraining air have been used for many years in recirculation systems and coating preparation. An ultrasonic centrifuge-type device has been used in paperboard mills to remove entrained air. High-shear dispersers such as a Rotor Stator-type mixing head are utilized in recirculation as well as in makedown. The latter, when used in coating recirculation, is very effective and probably is an under-utilized concept. The major operational problem attributed to mechanical air removal is the physical placement of the device within the system. If the system is well designed and maintained, it is probable that placement of the Rotor Stator mixer at the service tank would essentially eliminate air entrainment and prevent foam interference at the coater. Mills with a recirculation system utilizing this type of equipment have a much reduced need for chemical foam control agents. A Rotor Stator mixer operating continuously at the service tank has the potential to completely deaerate coating supply to the coater, providing a coating with a density essentially equal to the theoretical.

Foam Control Evaluation

General

The terms "antifoam" and "defoamer" are used interchangeably in the industry and quite often the question is asked for a distinction. "Defoamer" has come to be the term most used for all foam control agent applications. To the writer, the use of the term "antifoam" for foam control agent applied in coatings is more to the point because the need is to prevent air entrainment. The term "defoamer" as used in coating applications is a material used to "knockdown" foam that results from air entrainment. There are very few true defoamers available for paper coating due to FDA restrictions on the chemicals that can be added "neat" in coater recirculation without causing voids ("fisheyes" or "defoamer streaks") on the coated sheet.

Selecting the Most Effective Foam Control Agent

The literature contains much about the large array of chemicals used as foam control agents for coating, at one time or another, and much of this information is obsolete for two reasons:

1. FDA Requirements: Most mills request that the foam control agent for coating comply with "21 CFR 176.200, Defoaming Agents Used in Coatings," and seldom, if ever, is this distinction made in the literature.
2. Ineffective Chemical: The severe application conditions on the modern coater; i.e., coater speed, temperature, high foaming surfactants, etc., have made obsolete many formerly used materials.

These factors have substantially reduced the number of chemical types available to the foam control agent formulator.

Three major types of antifoam/defoamer compositions are available as foam control agents for coatings — water-based, oil-based, and polymeric.

Water-based products have been available for many years. The major advantage of this type is in coatings that are particularly susceptible to voids, "fisheyes." Cylinder board mills, using a furnish contaminated with wax or oil, and air-knife coating applicators, found this type of antifoam of particular value because the water-based products did not further aggravate the "fisheye" condition as could be anticipated with an oil-based product. The 1970's oil embargoes renewed interest in this type of composition, and newer products with more "punch" have been formulated for modern coatings.

Oil-based compositions are the most generally used compositions for coating applications because they seem to be the most adaptable and versatile, both for the chemical formulator and the user. Using paraffin oil as a base, the formulator adds hydrophobes as required depending upon the end use of the product, thereby custom fitting the composition to the coating formulator's needs.

The polymeric type product may be the composition of choice when the formulator does not want to use an oil-based composition and has found the water-based compositions lacking the "punch" required to control the foam. Compositions of this type can be very effective, particularly for knockdown, and are usually intermediate in their ability to deliver adequate carry-through.

The reputation of silicone-containing products, in contributing to glueability problems, right or wrong, has also restricted the formulator. It is the writer's opinion that silicone materials are very effective and reliable and definitely have a place in well formulated antifoams for all coating applications. It is unlikely there is a more effective antifoam, for long-term application, than a composition containing hydrophobic silica, a combination of silicone oil applied to a silica particle and subsequently heat treated.

Parameters of Foam Control Evaluation, Fritz and Hill *(2)*

"The degree of efficiency of a foam control agent for coating is dependent upon two equally important criteria:

1. The ability of the agent to control foam and/or entrained air in the coating system.
2. The effect of the agent upon the quality of the finished coated paper or paperboard product. Any meaningful evaluation of a foam control agent for a particular application must include tests which will satisfy both criteria."

"Foam Control Testing"

"Tests to demonstrate the defoaming ability of foam control agents are numerous and generally involve a mechanical means (shaker or agitator, etc.) to entrain air. The result of the test is determined by measuring the height of the foam head and/or by measuring the coating density. No specific test method is favored throughout the industry, probably due to the substantial variety of operations and the desire of the individuals involved to modify or develop a test which seemingly meets the particular mill's requirement.

There are general observations which can be made about foam control tests that are applicable to any test method that may be used:

1. It is important to limit the amount of foam control agent used to that which will adequately control foam but which will not so completely defoam the coating that functional differences between defoamers cannot be distinguished.
2. An air entrainment test using an agitator or blender should not be so violent as to be completely unrelated to machine conditions. The coating sample should be sized to allow sufficient agitation but, if the agitation is so severe that the foam control agent is not allowed to function, a meaningful test result is unlikely.
3. It is important that the coating to be evaluated be prepared as closely to the production standards as possible; all components, major and minor, should be included, and pH, solids, and viscosity should match machine conditions.
4. The coating sample used for each individual test should be as large as possible and practical. The physical movement of a larger sample within a test container more nearly represents mill conditions, is more difficult to stir violently, and reduces errors of sampling inherent in using smaller samples."

Specific Foam Control Tests

Much has been written about specific control tests for evaluating foam control agents, and it is likely that nearly all mills have their own specific and proven tests. For that reason, the writer does not intend to try to expand much upon what is already being practiced.

One general type of test used in various mills and by suppliers is probably the best evaluation tool for studying coating foam control agents. That test is the recirculating pump test utilizing a small surge tank and any type of pulp that will effectively recirculate a typical viscous coating color. This type of test has made obsolete the many mixer tests that have been used. The type of action given the coating by a recirculating test is much more typical of what happens on a coater than other tests, and a larger sample can be evaluated and studied for greater duration. A minor drawback is the larger sample size required (approximately 2–4 quarts), but that is also a strength of the test as indicated in the previous section, "Foam Control Testing." A key to constructing a satisfactory test apparatus is selection of the recirculating pump. The Wilden M-1 double-diaphragm air pump constructed of polypropylene is ideal for this application. This pump is light in weight, relatively inexpensive, and will pump all coating colors regardless of viscosity.

Observation of Side Effects

Any thorough foam control evaluation using antifoams is going to result in testing products where coating voids will be identified as a problem for one or more of the products tested or at certain addition levels for all products tested. With a recirculating tester, the problem can sometimes be identified by dipping a polished, stainless steel spatula into the coating and, as the coating drains away, observing for tendency to "fisheye." This technique can become a part of the test and repeated several times to determine whether or not, as the coating is being recirculated, the "fisheyes" disappear in time.

Following the recirculating test, drawdowns can be run with the test coating on uncoated basestock or on clear vinyl sheets as a replacement for glass plates (equally as effective as glass, but storable for future reference). The writer prefers uncoated basestock because the vinyl or glass introduces a meaningless parameter and, in the case of a recycled furnish, the basestock may present conditions more severe than glass. Other tests can also be run on the test coating as desired.

Literature Cited

1. Busch, T. W., *Paper coating additives,* TAPPI PRESS, 1978, pp. 5–6.
2. Fritz, J. H. and Hill, J. A., *Paper coating additives,* TAPPI PRESS, 1978, pp. 47–48.

Bibliography

Becher, P., *Emulsions: Theory and Practice,* A.C.S., Reinhold, NY, 1957, Monograph Series.

Casey, J. P., *Pulp and Paper,* Vol. III, Interscience, New York. 1966, pp. 1674–77.

The Institute of Paper Chemistry, *Defoamers — An Annotated Bibliography,* July 10, 1973.

Gunning, J. R., *Pulp and Paper Manufacture,* Vol II, McGraw Hill, New, York, 1969, pp. 475–87.

Lichtman, I., "Defoaming Agents" in *Chemical and Process Technology Encyclopedia,* McGraw Hill, New York.

Nadleman, A. H. and Baldauf, G. H., "Coating Formulations: Principals and Practice" in *Lockwood Trade Journal,* Lockwood Trade Journal Co., New York, 1966.

Ross, S., "The Inhibition of Foaming" in *Engineering and Science.* Series 63, Rensélaer Polytechnic Institute, Troy, NY, 1950.

Thorkildsen, R. E. and Rich, S. R., "Centrifugal-Sonic Defoaming of Coatings," *Tappi* 49(9): (1966).

CHAPTER 4: WATER RETENTION AND RHEOLOGY MODIFIERS

Teng-Shau Young and Charles Lee Burdick

Introduction

Water retention and rheology modifiers (WRRMs) are used widely in paper coating formulations. This class of additives, sometimes called "thickeners," are added to coatings primarily to provide two functions, water retention and rheology control. These functions are known to play an important role in determining the runnability, coat weight, and optical/mechanical properties of the finished coated paper or paperboard. Therefore, the proper use of such an additive, to be referred to as WRRM throughout this chapter, can be critical to the success of coated paper production.

WRRMs are generically water-soluble or water-swellable polymers of moderately high molecular weights (typically over 10,000 in daltons). Each of them can be considered to consist of a long polymeric chain known as the "backbone," to which a large number of hydrophilic functional groups are chemically attached. The functional groups, which contribute to water affinity of the polymer, are generally either polar nonionic, such as hydroxyl or ether, or ionic, such as salts of carboxylic acid. Some of the well known WRRMs are sodium alginate and carboxymethylcellulose (CMC).

By virtue of their high water affinity and large hydrodynamic size in aqueous media, WRRMs can substantially "thicken" the aqueous coating vehicle and hence the coating color at a relatively low dosage. The role of a WRRM in the coating color, however, often goes far beyond merely thickening the color. For instance, some WRRMs can alter the structure of a coating through interaction with the pigment(s) and/or the latex binder *(1–3)*. Appreciable changes in coating rheology and consolidated coating structure consequently result. Thus, while the WRRM only represents a minor proportion of a typical paper coating, its overall influence on the coating properties can be very significant.

Many of the changes in coating structure and rheology brought about by a WRRM can directly impact coating processability and end properties of the coated paper. The viscosity enhancement of the coating vehicle hinders the movement of water into the paper substrate, and is largely responsible for reduced water loss on the coater. This leads to better controlled coating solids before blade or rod metering, thereby minimizing runnability problems associated with excessively high solids. The viscosity enhancement of the coating vehicle also limits liquid and coating penetration during and after the metering action, which leads to reduced binder migration and enhanced coating holdout. The development of coating structure by some WRRMs through association with pigment/latex can also contribute to water retention, control of binder migration, and enhanced coating holdout.

Virtually all of the rheology-modifying water-soluble polymers can also provide some pigment/fiber binding function. Some of these polymers, including starch, polyvinyl alcohol, and proteins, are in fact used in coating formulations primarily as binders or co-binders. These polymers are described in another TAPPI monograph on coating binders, and therefore will not be discussed here. The present chapter is concerned only with additives whose primary intended function is for water retention and rheology control, such as alginate and CMC.

A variety of WRRMs are commercially available. These WRRMs can lead to a wide range of coating properties, depending on the chemical structure and molecular size. In selecting a WRRM for a particular coating process, a paper coating practitioner seeks to attain the best combination of flow behavior, water retention, coat weight control, holdout, and binder migration. The selection of a particular WRRM over others usually depends on the specific coating properties needed for the concerned paper grade and coating process. Economics, ease of use, and supplier reliability are also factors in the decision making.

This chapter gives an overview of WRRMs of use in paper coatings and the underlying mechanisms by which they function. Commercial WRRMs covered herein include alginates, CMCs, hydroxyethylcelluloses (HECs), and alkali-soluble or swellable polyacrylates. All of these products have seen significant use in the industry. A new class of WRRMs, called associative thickeners, is also briefly described for their potential usefulness in high-speed coating processes. Several other additives that can affect the flow properties of coating colors, but are not WRRMs by definition, are also briefly mentioned at the end of this chapter.

Prior to discussion on these individual products, a general review of current theories on water loss, water retention, and rheology modification is presented. Test methods that have been reported in literature for the evaluation of water retention and rheological properties of paper coatings are also briefly described.

Current Theory and Test Methods

Theory — Water Retention

The mechanisms of on-machine water loss have been studied extensively for years. The understanding of the water migration phenomenon was significantly advanced as a result of the early work of Clark, Windle, and Beazley *(4)*. Since then, further understanding has been brought about by Eklund and his colleagues *(5, 6)*, Baumeister *(7)*, Engström and Rigdahl *(8)*, Fujiwara *et al. (9)*, and others. In this section, water loss mechanisms that have appeared in the literature are briefly reviewed. The role of WRRMs in controlling water migration is also discussed.

Water Loss Mechanisms

Loss of water from the coating color to the web generally involves migration of water at the coating/paper interface into paper substrate, followed by motion of water in the bulk of the coating to the interface. At least three different types of driving forces have been proposed for migration of water from the interface into paper substrate. These driving forces are (a) capillary pressure, which drives water into paper substrate *(5)*, (b) mechanical pressure encountered on the coating applicator and at the entrance of the metering blade or rod, and (c) vacuum generated under the coating, as a result of compression of the paper stock under the blade and recovery of the compressed stock after the blade; this causes a "sponge" effect, i.e., vacuum-driven water migration after the blade *(9)*.

Water loss due to capillary pressure occurs at all times after the coating has been applied onto the paper web. It is the primary source of water loss during the dwell period. However, it accounts for only a small portion of the total water loss at completion of the metering action if pressure migration also occurs in the concerned process *(5)*. This is particularly the case in high-speed coating processes using a short-dwell coater, where the time of coating/web contact is short.

Pressure-driven water migration (or pressure migration) at the applicator is most noticeable with a flooded-nip applicator where the coating is subject to a large positive pressure entering the nip *(5)*. This type of water loss is not significant in coating processes involving a low-pressure applicator such as a short-dwell or puddle applicator.

Pressure-driven water migration also occurs to a significant degree under the blade, where a large pressure exists. Liquid loss under the blade has been shown to be significant with low solids coating colors in a recent study *(9)*. In high-speed coating of a high solids color, however, a filter cake may form at the coating/paper interface after losing a relatively small amount of water *(7–9)*. Once the cake is formed, subsequent water loss is substantially reduced. Such high solids coatings have been shown to lose water largely after the blade, as a result of the "sponge effect" described earlier.

Water loss before and after the doctoring blade or rod can both impact the effectiveness of a coating operation. The need for adequate water retention before the blade/rod is obvious for its direct impact on runnability. Some measure to control water migration after metering is also desired, as excessive water migration may cause binder loss/migration and loss of web strength. While the effectiveness of WRRMs in reducing both the before and after-blade water loss has been known for years, the exact mechanisms by which water retention is achieved are not completely understood. Some of the mechanisms most likely to be responsible for on-coater water retention are discussed below.

Water Retention Mechanisms

Water migration driven by either capillary pressure or mechanical pressure can be significantly slowed by increasing the viscosity of the coating liquid phase. The increased liquid viscosity also inhibits migration of the liquid from the bulk to the paper surface. Thus, the most obvious and widely accepted mechanism for water retention involves viscosity enhancement of the liquid phase through the addition of a WRRM. To maximize the liquid viscosity, WRRMs commonly used for this purpose are not strongly adsorbed to coating pigments. The use level of such water retention agents is usually determined by the viscosity of the resulting coating.

Water retention is also affected by factors other than aqueous phase viscosity. The state of the wet coating structure has been suggested to influence the dehydration rate of coatings *(10)*. That is, different degrees of pigment flocculation may result in different flow pathways for the coating vehicle, and hence varied dehydration rates. The phenomenon of cake formation at the coating/web interface as described earlier can also play a role *(7–9)*. This mechanism is believed to be important in high-speed, high solids coating applications where a relatively small amount of water loss can lead to immobilization of the coating solids at the coating/paper interface. To achieve this type of water retention, the WRRM needs to induce coating immobilization at a relatively low solids level. Some of the more clay adsorbing WRRMs such as hydroxyethylcellulose, may attain water retention partly through cake formation.

It should be kept in mind that the water retention WRRMs are also rheology modifiers which can alter the properties of both wet and consolidated coatings. When selecting a WRRM for water retention purpose, its potential impact on coating rheology and coated paper properties must also be considered.

Theory — Rheology Modification

WRRMs commonly used for water loss control also significantly change the rheological behavior of a coating color. This nature, in fact, is being utilized to allow control of certain rheology-related coating properties. For instance, use of a proper WRRM can facilitate the control of high-shear coating rheology, and hence the coat weight. Its use also normally leads to low-shear rheology desirable for mixing, pumping, and recirculation of the coating color. The coating structure after blade or rod doctoring, which is believed to affect coating holdout, can also be strongly influenced by the WRRM.

A WRRM not only increases the viscosity of the coating through liquid thickening, but also alters the state of pigment or latex flocculation. Since the state of flocculation can strongly influence the rheology and wet/dry structure of the coating color, WRRMs usually play an important role in determining coating processability and end properties. How and to what a degree a WRRM affects the coating properties depend primarily on its interaction with pigments and/or latex and its hydrodynamic size in solution.

Commercially used WRRMs vary in molecular weight (MW) over a wide range (10,000 to over 1,000,000). Their chemical natures also differ substantially, leading to distinct associative behavior in typical paper coatings. Virtually all the WRRMs used in North America today are either nonionic or anionic in nature. The anionic WRRMs generally adsorb less to kaolin clays than the nonionic WRRMs due to the repulsive force between the WRRM and the negatively charged clay surface *(1, 2, 11)*. The extent of clay adsorption has been shown to increase with increased MW for cellulosic WRRMs *(2)*.

WRRMs of lower MWs (i.e., less than 100,000), in general, cause a lower degree of pigment or latex flocculation than the high MW ones *(1, 2)*. This phenomenon has been attributed to two properties of low MW polymers: (a) relatively low clay adsorption, and (b) lower probability of causing bridging flocculation of clay *(3)* or depletion flocculation of latex *(12, 13)*. As a result, coatings thickened with a low MW WRRM are relatively well dispersed. Some of the low MW WRRMs have been observed to behave as protective colloids, deflocculating the clay when under shear *(1, 2)*.

Under low shear conditions, addition of a low MW WRRM leads to increased coating structure, but to a relatively small degree *(2)*. Under high shear conditions, low MW WRRMs generally give rise to more dispersed coatings, even at relatively high coating solids. Good high-speed runnability is therefore attainable. The resulting coating usually is less pseudoplastic (less shear thinning) compared to a coating containing a high MW thickener of the same chemical nature and at the same solids and Brookfield viscosity. This property arises from the fact that the coating containing a low MW WRRM lacks structure under low shear conditions, and hence requires a higher concentration to achieve the same Brookfield viscosity *(2)*. On the coater, this can translate into a relatively high coat weight. This effect is sometimes offset by the generally strong water holding of the low MW WRRMs, which reduces the change in coating solids and coating viscosity under the doctoring blade or rod.

In contrast, high MW WRRMs tend to cause more coating flocculation by virtue of their larger hydrodynamic size and higher tendency to adsorb on clays *(1, 2)*. The flocculation effect is particularly profound with strongly clay-adsorbing polymers such as polyvinyl alcohol, HEC, and some proteins *(14, 15)*. The higher degree of flocculation often results in a stronger coating structure and a higher coating viscosity at low shear rates. Higher thickening efficiency and better coating holdout can thus result.

When operating within a carefully selected solids and speed range, coatings thickened with moderately high MW WRRMs are generally quite shear thinning. The floc structure which exists at low shear rates is largely destroyed by mechanical shear forces encountered during mixing, pumping, coating application, and doctoring. This can lead to a relatively low coat weight which is supposedly desirable in light-weight coated (LWC) paper making. However, special caution must be taken to avoid too strong a floc structure. If the floc structure is excessive, it may not be broken down quickly enough under the blade by the imposed shear force. A high flow resistance can thus arise, leading to blade instability and accompanying runnability problems. Such a problem is more likely to occur when the coating solids is high (strong floc structure and high flow resistance) or the coating speed is high (very little time allowed for doctoring action, on the order of 10^{-4} seconds or shorter).

The runnability concerns constitute the major restriction in using high MW WRRMs. Normally, the upper limit on the WRRM MW is dependent on the coating solids, clay-adsorbing nature of the WRRM, and coating speeds. A WRRM of a higher MW may be used if it is less prone to clay adsorption and/or the coating solids is relatively low. To ensure acceptable runnability, careful laboratory and coater trials are warranted prior to the use of a high MW WRRM.

The selection of a WRRM can be aided by laboratory tests that allow an assessment of the influence of different WRRMs on coating water retention and rheology. A brief review of these test methods follows.

Test Methods — Water Retention

Despite the numerous methods reported in the literature, there is not a universally accepted method for water retention measurement. Among the existing methods, the S.D. Warren method which is based on conductivity measurement *(16)* is most commonly practiced. This test is conducted at atmospheric pressure under static conditions. Although sensitive to the conductivity of the aqueous coating vehicle, the water retention times obtained usually correlate to the aqueous phase viscosity. However, this method does not simulate water migration behavior under high-pressure, dynamic conditions, and fails to predict coater runnability on some occasions.

A few other methods also operate under atmospheric pressure. Some involve the measurement of changes in gloss of the dehydrating coating. The dehydration rate measured is usually related to the time required for the coating to reach its gloss immobilization point (GIMP) *(17)*. For example, the method reported by Beck *et al.* involves application of the coating onto an unglazed ceramic plate and monitoring of the light reflected from the dewatering coating until the coating reaches its GIMP *(18)*. Another method not based on optical measurements was reported by Taylor and Dill *(19)*. Their method involves the measurement of the speed of sound waves traveling through a paper sheet. Water drainage into the paper is signified by a decrease in sound velocity due to fiber swelling. All of the above methods suffer from the same problem as the S. D. Warren method, i.e., inability to simulate pressure migration under dynamic conditions.

A number of methods to study water migration under pressure have been reported in the literature. In these methods, the porous substrate can be either paper or a membrane filter, and the driving force for liquid migration can be either positive pressure exerted by oil or air, or vacuum. Several involve the measurement of the rate of change in coating gloss *(4, 20)*. The methods reported by Beck *et al. (18)* and Sandås *et al. (6)* are based on pressure filtration, in which the amount of water penetrating through a filter into an absorbing paper is determined gravimetrically. Somer *(21)* described a film splitting method in which the time required to reach a solids level at which the amount of water is just enough to fill the voids in the coating is used as a measure of dehydration rate. Despite the use of pressure in these methods, none of them has shown a definitive correlation with pilot coater or mill data. The lack of such a correlation was in part a consequence of insufficient pilot coater or mill data.

A dynamic water retention test method has been described by Jones and Heterington *(22)*. The test involves the use of a puddle-type laboratory blade coater. The increase in coating solids with time is used as a measure of dehydration rate. This approach has also been used with a pilot coater to study water retention of paper board formulations *(23)*. Some of the more definitive water loss data have also been obtained using pilot coaters. Engström and Rigdahl *(8)* studied water migration by scraping and analyzing the coating at varied points before the dryer. The same technique has also been used by Fujiwara *et al. (9)* with a high-speed pilot coater.

Test Methods — Coating Rheology

The most commonly used rheological test involves the use of a Hercules Hi-Shear Rheometer *(24)*. The rheometer has a bob-and-cup geometry and can be operated at shear rates up to 100,000 s^{-1}. Useful information such as high shear viscosity, pseudoplasticity, and coating thixotropy can be obtained from this test. The shape and smoothness of the thixotropic curves also allow early detection of potential runnability problems. Major deficiencies of this rheometer are its lack of temperature control and slow shear rate acceleration. About 20 seconds are required for the rheometer to reach its maximal rotating speed; this is much longer than the time typically involved in coating doctoring.

Rheological properties at even higher shear rates (up to 10^7 s^{-1}) and shorter deformation time may be attained from other rheometers that involve high speed, transient flow of coating through a short capillary *(2, 25)*, slit-die *(26)*, or annular gap *(27, 28)*. Such rheometers may operate under controlled stress (using positive pressure) or controlled strain rate (using a displacement pump). Despite a generally larger coating consumption and more tedious operation, such methods allow better temperature control and simulation of rapid, instant shear of coating under a blade or rod. The instrument described by Ramthun *et al. (28)* appears to be quite useful because (a) it allows continuous mixing and pumping of the coating color prior to the high shear measurement, which better simulates machine conditions, (b) shear rates up to 10^7 s^{-1} are obtainable, and (c) it is relatively easy to adjust the gap and speed, and hence shear conditions.

Useful information on coating rheology under low shear conditions can be obtained from a simple Brookfield viscometer. Viscosities measured at 100 and 20 rpm are widely used by mill and laboratory personnel to quantify the low shear strength of the coating.

In the mills, the coating is generally thickened to a target Brookfield viscosity which is determined empirically.

Coating rheology at low-to-moderate shear rates can be quantitatively measured using a general purpose rheometer. Such a rheometer normally can operate under steady shear, oscillatory shear, and/or transient shear conditions under either controlled strain rate or stress. By properly selecting the test conditions, information on structural breakdown, coating viscoelasticity, degree of coating flocculation, and after-shear coating strength can be obtained *(2)*. Commercial rheometers suitable for such studies are available from Rheometrics, Bohlin Reologi, Haake, Sangamo, Contraves, and others.

Commercial Water Retention and Rheology Modifiers

Only three types of WRRMs are of significant commercial use in paper coating industry. They are the alginates, cellulosics (e.g., CMCs and HECs), and polyacrylates. This section describes their chemical structure, properties, and commercial practice in paper coating applications. A brief discussion on associative WRRMs is also included.

Alginates

Algin is a natural polysaccharide which is found in brown seaweeds of the group "Phaeophycea" *(29)*. These seaweeds are harvested in several coastal locations around the world and chemically processed to yield the various grades of alginate useful for paper coatings and other applications.

Alginate polymers have found prevalent use in size press coatings and are also employed to a large extent as WRRMs for bleached board coating formulations. Despite a relatively high price compared to other paper coating WRRMs, alginates display desirable performance characteristics that have maintained their presence in the paper coatings industry.

Structure

Sodium alginate is a linear anionic polymer consisting of units of guluronic acid sodium salt and mannuronic acid sodium salt residues as shown in **Fig. 4.1**. It is reported *(30)* that the alginate polymer consists of segments of polyguluronic acid salt block polymer, polymannuronic acid salt block polymer, and also segments of polyguluronic/mannuronic acid alternating polymer as shown in the figure.

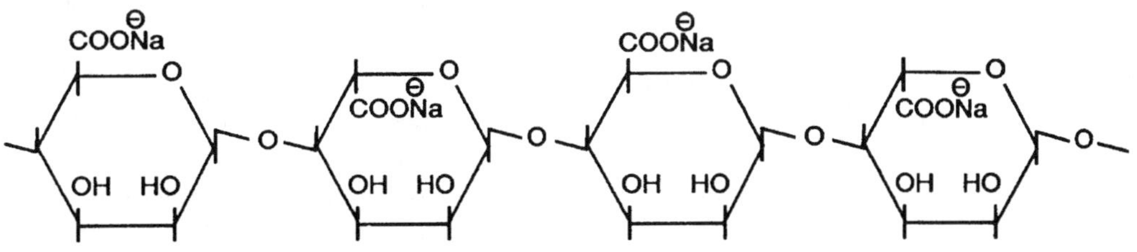

Fig. 4.1. Idealized structure of sodium alginate

The guluronic acid salt and the mannuronic acid salt residues present in alginate polymers are known to associate with multivalent ions, such as dissolved calcium, and this can affect the viscosity and rheology of this polymer in solution *(31)*. For this reason, alginates are often used in paper coatings and other applications in combination with chelating agents, such as polyvalent phosphate salts, which complex calcium and give more predictable performance of the alginate *(32)*.

The viscosity of sodium alginate at a given concentration is a function of the length of the polymer chain, known as the degree of polymerization and is also related to the ratios of guluronic and mannuronic acid residues of the polymer *(31, 32)*.

Properties

One of the properties of alginates of importance to paper coatings applications is its ability to promote water retention. Solutions of sodium alginate in water are known to exhibit relatively low penetration into porous substrates. This makes alginate especially effective for use in size press applications for holdout of starch solutions on the surface of the paper. When employed in paper coatings, alginate similarly tends to hold the coating's pigments and binders on the paper surface.

The high water retention properties of alginate are well suited for bleached board coatings which typically are applied at relatively low speeds and high solids. Additional benefits associated with alginate in bleached board include good binder holdout which can improve glueability of the coated board, and good film-forming properties which can improve the board's printing properties.

Common Practice and Methods of Use

Sodium alginate is usually employed as a granular solid. The product may either be predissolved in water and then added in solution form to the base coating, or more typically, the granular sodium alginate is added directly to the paper coating during batch makeup and allowed to dissolve in situ. The dispersibility of the granular sodium alginate appears readily amenable to the batch process of bleached board coatings makeup.

Various viscosity types of sodium alginate are commercially available *(32)*. The selection of the alginate viscosity type for any given coating application is based upon an analysis of the relative cost, ease of use, and performance criteria. The higher viscosity types of alginate are the most dosage effective. However, coating viscosity control with higher viscosity WRRM types is generally more difficult, since small changes in WRRM can result in steep coating viscosity response.

Cellulosics

Cellulose has a number of uses in addition to being the major component of paper. The cellulose raw material can be converted by chemical processes into water-soluble polymers useful for rheology control of paper coatings. While cellulose itself is insoluble in water, carboxymethylation or hydroxyethylation of cellulose yields the water-soluble derivatives known as "CMC" and "HEC," respectively. These derivatives of cellulose have found extensive use as paper coating WRRMs.

Other cellulose derivatives such as methylcellulose, hydroxypropylcellulose, and methylhydroxypropylcellulose can be produced with chemical modification *(33)*. At present, none of these is known to have significant commercial use in paper coatings. Another relatively new class of cellulosic derivatives, namely hydrophobically-modified cellulosic polymers, is covered in more detail in a later section on associative WRRMs.

Carboxymethylcellulose (CMC)

Structure

Either pulp or cotton linters can be employed as a raw material for the production of CMC. To produce CMC, a chemical process is employed to convert a portion of the hydroxyl groups of the cellulose molecule to sodium carboxymethyl groups. The resultant polymer is an anionic linear polysaccharide similar in some respects to sodium alginate but with a different stereochemistry. A model structure of CMC compared to the starting cellulose raw material is shown in **Fig. 4.2.**

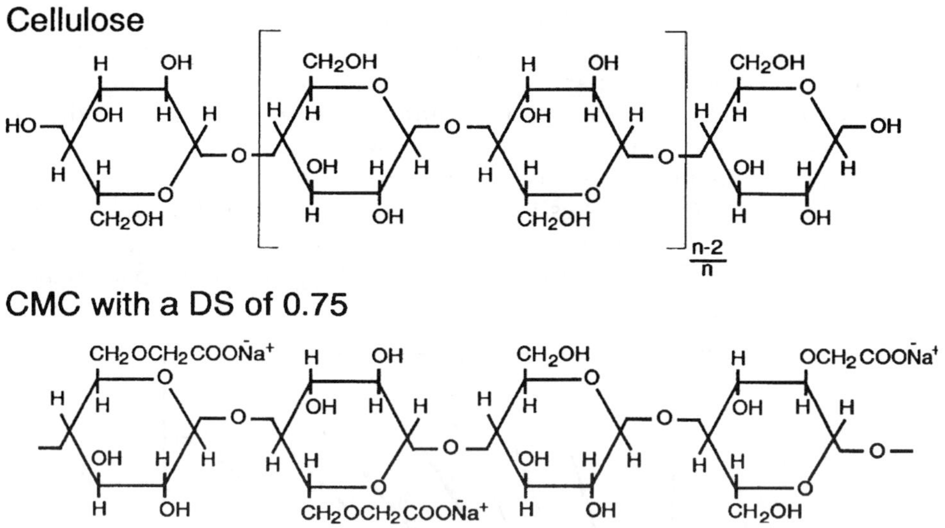

Fig. 4.2. Structure of CMC compared to cellulose

Because the cellulose raw material employed for CMC synthesis has many hydroxyl sites available for reaction, the CMC final product is a mixture of many molecular species. Some of the individual glucose units of the cellulose chain may become derivatized with more than one carboxyl group, while other units remain unreacted. It is a common practice to describe the degree of reaction of the carboxylation as the "DS" or degree of substitution of the final polymer. For typical CMC employed in paper coatings, the DS is usually O.6 to 0.9, which means that for every 10 glucose units in the cellulose chain there are 6 to 9 carboxymethyl functional groups.

At DS values lower than this range, a proportion of the CMC may not be completely soluble in water, resulting in a phenomenon known as "granularity." Granularity is evidenced by the presence of swollen CMC gel particles that do not dissolve in water even after prolonged agitation. This phenomenon can be undesirable in coating practice and suppliers of CMC have attempted to keep granularity of their products to a minimum.

Properties

CMC forms stable solutions over a broad pH range. It is more tolerant of calcium ions than sodium alginate and does not require chelating agents for its use in paper coatings. As alginates, CMCs are also good film-forming agents and function as co-binders in paper coatings. The water retention performance of CMC in paper coatings is to some extent formulation dependent, but is generally very good.

Usually, low to medium viscosity-type CMCs tend to give a higher S. D. Warren water retention time compared to a high viscosity type. The longer water retention time observed for the lower viscosity CMC is due in part to the higher dosage employed. Another reason is its relatively low clay adsorption compared to that of a high viscosity-type CMC. Both factors cause a higher polymer concentration in the aqueous phase of the coating, which leads to enhanced water retention.

Common Practice and Methods of Use

CMC is the predominant WRRM employed industrially in free sheet paper coating formulations. It is also employed in some bleached board and recycled board coatings. CMC has found little utilization in LWC papers due to the use of starch as the primary binder/thickener. There also appears to be a preference in this industry segment for liquid WRRMs over dry powdered WRRMs such as CMC.

CMC products are supplied in either fine powder or coarse granular forms. Most commonly, CMC is employed in paper coatings by first predissolving in water and then adding this solution to the coating color in sufficient quantity to achieve the targeted viscosity and solids. The coarse granular types of CMC can also be added directly to the coating formulation and allowed to dissolve in situ.

A broad range of CMC viscosities is commercially available (**Fig. 4.3**). Industrially, the most prevalent use of CMC is with the low viscosity types. There is also, however, a fair amount of use of medium molecular weight CMC types to meet particular coating performance requirements.

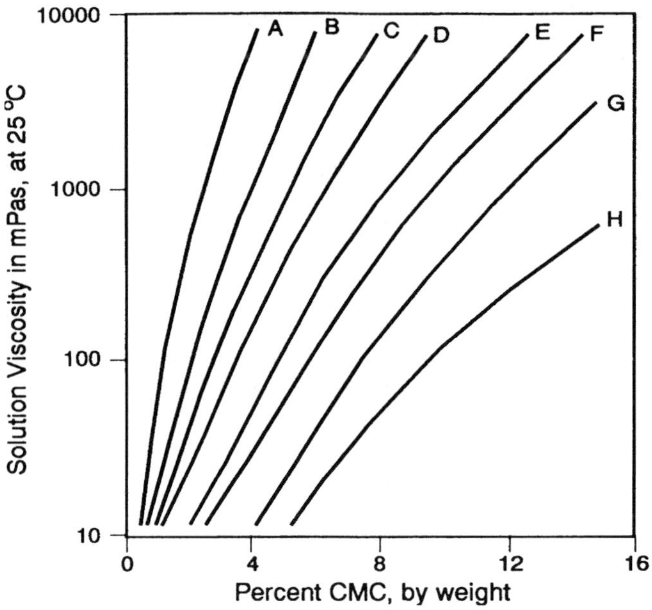

Fig. 4.3. Range of CMC viscosity types available

Low viscosity types of CMC are often employed in higher solids coating formulations. The medium to high viscosity CMC types generally find use in coating formulations with somewhat lower solids, corresponding to a higher proportion of the aqueous phase. This allows better dosage control of the higher molecular weight WRRM to achieve the target coating viscosity.

Hydroxyethylcellulose (HEC)

Structure

Hydroxyethylcellulose is a nonionic water-soluble derivative of cellulose. This WRRM is the reaction product of ethylene oxide with cellulose. A representative view of HEC is shown in **Fig. 4.4**. The quantity of ethylene oxide that is reacted with the cellulose in production of HEC is expressed as molecular substitution or "MS." Typically the MS of HEC is in the range of 2.0 to 3.0 moles of ethylene oxide per glucose unit of the cellulose chain.

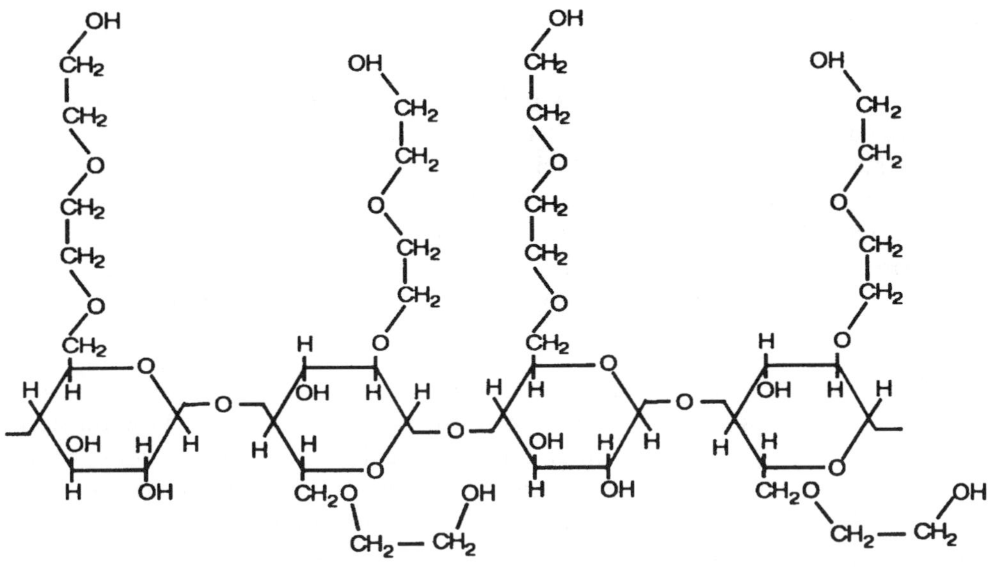

Fig. 4.4. Idealized structure of hydroxyethylcellulose

Properties

There are some essential differences between HEC and CMC. HEC has been found to adsorb to kaolin clays to a greater extent than the corresponding molecular weight grade of CMC *(2)*. This is associated with correspondingly lower S. D. Warren water retention values of some paper coatings thickened with HEC compared to coatings thickened with the equivalent dosage of an equal viscosity type of CMC. In those cases where higher water retention is desired, a greater dosage of a lower viscosity-type HEC is often employed.

HEC has been reported to give higher K&N ink absorption in certain paper coating formulations compared to CMC *(14)*. This has been found to be desirable in some board grades and rotogravure paper coating formulations. It has been proposed that mild pigment flocculation induced by HEC is responsible for the improved ink receptivity *(14)*.

Because HEC is a nonionic water-soluble polymer, it is not adversely affected by dissolved ionic species such as calcium and ferric ions present in the coating color. This allows greater predictability of performance and latitude to coating practitioners when variable water quality conditions may be encountered.

Common Practice and Methods of Use

HEC is employed industrially in free sheet, bleached board, and recycled board coating applications. There are a broad variety of HEC viscosity types commercially available varying from low to high viscosity types, similar to the range available with CMC. The full spectrum of these HEC viscosity types is employed in various paper coating formulations.

HEC is available in surface-treated grades that allow for easy dispersion and dissolution without lump formation. This can be a distinct handling advantage over other dry WRRMs, particularly with use of the higher viscosity types. HEC is also commercially available in a high-solids fluid suspension form that may offer further handling advantages to some end users.

Polyacrylates

Structure

Polyacrylates are the only petroleum-derived synthetic WRRMs that have found significant use in paper coating. In general, they are alkali-soluble or swellable polymers consisting of acrylic acid or its salt, which provides the water affinity. Commercial products are usually copolymers containing acrylate (e.g., ethyl acrylate) and acrylic acid (or methacrylic acid). Some of the alkali-swellable polyacrylates are lightly crosslinked and are in a form of "microgels" *(34)*. Hygroscopicity and viscosifying power may vary widely depending on the chemical composition and size of the polymer *(35)*.

Properties

Acrylate-based WRRMs can be manufactured in solid or emulsion form, but only the emulsions have found significant commercial use in paper coatings. The emulsions are normally acidic. Under such conditions, the acrylic acid or methacrylic acid groups are not dissociated, rendering it possible to insolubilize the polymer in aqueous media. The emulsion can be converted into a clear, viscous solution upon dilution and addition of a base, as typically practiced in coating preparation. This phase change is a result of ionization of the acrylic/methacrylic acid groups under basic conditions which causes the polymer chains to disperse into the aqueous solvent. At a sufficiently high pH, the polymer chains are highly extended, owing to intra-molecular charge repulsion, or the "polyelectrolyte effect" *(36)*. This effect accounts for a substantial part of polyacrylate's viscosifying power.

The water solubility of alkali-swellable polyacrylates is dependent on the pH. Their solubility is generally low under acidic conditions, since many of the acrylic acid groups are not dissociated. The pKa values, which govern the dissociation of acrylic acid groups, may vary significantly with the polymer composition. However, all the polyacrylates known to be used in paper coatings are soluble at a pH above 8.

The alkali-swellable nature of polyacrylates allows them to be supplied and stored as liquid emulsions, which are more convenient to handle than a dry product. The liquid-handling convenience is most appreciated in mills that are equipped with automated coating kitchens, and has been a major factor in bringing polyacrylate emulsions into some of the LWC mills.

The majority of the commercial polyacrylates used in paper coatings are believed to have an MW of 100,000 or higher. Because these WRRMs carry anionic charge under basic conditions, most of them are not strongly adsorbed to kaolin clays. In this respect, polyacrylates are similar to CMCs of relatively high molecular weights. Commercially used polyacrylates generally have a good thickening efficiency due to their relatively high MW and extended chain conformation. Their solution properties normally are not sensitive to temperature.

Because of the role of alkali-swelling and polyelectrolyte effect in their viscosifying mechanisms, solution properties of polyacrylates are highly sensitive to pH and salt content of the coating vehicle. Consequently, rigorous control of coating pH is essential when a polyacrylate WRRM is used. Caution should also be taken to avoid the presence of extraneous multi-valent ions in the coating, which may cause polymer precipitation and loss of efficiency.

Common Practice and Methods of Use

Polyacrylate emulsions are used commercially in both coated paper and paperboard production. They are commonly used at a dosage level ranging from 0.15 to 1.0 parts per hundred parts pigment, depending on the coating compositions and solids, and the type of WRRM used. The resulting coating is usually shear thinning and fluid.

The emulsion is usually added last to the coating color to thicken the coating to the target Brookfield viscosity. A predetermined amount of base, such as ammonium hydroxide, is added to the base coating to ensure basicity. The pH of the resulting coating usually is designed to fall between 8 and 9.

Associative Thickeners

The term "associative thickener" is relatively new to the paper coating industry. In fact, it was borrowed from the water-borne paint industry where thickeners that show significant associative behavior in latex-based coatings are widely used. It was suggested by E. Glass in 1988 *(37)* that some of the synthetic associative thickeners commercially used in latex paints may also be useful in paper coatings. That suggestion, however, was primarily based on previous experience with latex paints. To date, supporting paper coating data have not yet appeared in the literature. Recently, another type of associative thickener based on cellulosic derivatives has been developed and tested for paper coating applications *(2)*. The associative cellulosic thickener (ACT) has been shown to display unique associative behavior and offer some properties desirable in high-speed coating processes. No report of commercial use of any of the associative thickeners has yet appeared in public domain, but their interesting properties and potential application in high-speed coating processes have drawn substantial interest from paper coating researchers. This section briefly describes the nature and properties of both types of associative thickeners.

All the associative thickeners referred to herein are hydrophobically modified. The substitution of hydrophobic moieties on a water-soluble polymer gives rise to a unique property in aqueous media. That is, the dissolved polymer molecules become associated with one another in solution through hydrophobe-hydrophobe bonding. This type of inter-molecular association can result in an increased hydrodynamic size for the polymer, thereby leading to an increased solution viscosity *(2, 37)*. Thus, a high thickening efficiency may be obtained from a polymer of a relatively short chain length. An important feature of the hydrophobe association is its weak bonding force, which is of the London-van der Waals type. As a result, solution or suspension structure induced by the hydrophobe association can be broken down with less energy. This is of practical importance in paper coating operations because it allows metering a highly structured coating at high speeds.

The synthetic associative thickeners discussed by Glass can be divided into two classes. One class comprises of a variety of hydrophobe ethoxylated urethanes (HEUR); the other includes hydrophobe modified alkali-swellable emulsions (HASE).

The HEUR products known to the latex paint industry are generally of the low MW type. They commonly contain hydrophobes at the ends of a hydrophilic polymer backbone. Association of such polymer chains leads to longer, linear chains. HEURs usually associate with the latices used in latex paints, a property largely responsible for their performance in paint formulations. Latex coatings thickened with HEURs are known to give good "leveling," a desired property in paint applications *(37)*.

Their usefulness in paper coatings is not so clear. Some of their unique rheological properties observed in latex paints may not be as visible in paper coatings, where the latex concentration is much lower. The leveling property imparted by HEUR products may actually be a disadvantage, as it may be allied with poor coating holdout. Some improvement in the thickening efficiency probably will also be needed to be cost effective.

HASE products are basically hydrophobically modified polyacrylates *(38)*. Like most polyacrylates, they are not strongly adsorbed to clays. Because of hydrophobe association and their relatively high MW, HASE products probably will give reasonably good thickening efficiency in most paper coatings. Like other synthetic WRRMs, the holdout property of the paper coating may be a concern.

The associative cellulosic thickeners (ACTs) are hydrophobically modified products of cellulose ethers. ACTs that have been tested so far for paper coating applications are primarily hydrophobically modified HECs. In addition to inter-polymer hydrophobe association, the ACTs have been found to also associate with clays and styrene-butadiene latex *(2)*. The unique associative behavior has been shown to give rise to a fluid coating under high shear stress, but a strong coating structure after mechanical shear. Therefore, ACTs represent a new class of WRRMs that may attain both good coating holdout and high-speed runnability. ACTs are believed most suitable for LWC formulations owing to their desirable high shear and holdout properties.

Effect of Other Additives

While the WRRM is the primary additive which controls the coating rheology, additives used for other purposes may also affect the flow properties of the coating. Additives that can cause an appreciable change in coating rheology include dispersants, defoamers, and insolubilizers. Since the functions and properties of these additives are described in details in separate chapters, only a brief discussion of their possible effects on coating rheology is given here.

The effect of a dispersant on coating flow properties is obvious. When used at an appropriate level, a dispersant helps defloc the pigment and therefore gives reduced coating viscosity. However, using an excessive amount of dispersant can instead cause an increase in coating viscosity. In addition, a dispersant may influence the adsorption of a WRRM on clay *(10)*, and hence the rheology of clay-containing coatings.

Defoamers are used widely to prevent formation of excessive foam, which can sometimes cause problems in applying or doctoring the coating color. It is well known that entrapment of fine air bubbles may influence the viscosity of a suspension, such as a paper coating. Thus, a defoamer may affect the flow properties through its influence on the amount and/or size of the entrapped bubbles. A defoamer may also interact with certain coating ingredients. Therefore, compatibility with other coating components needs to be considered in defoamer selection.

Some of the insolubilizers for starch, soy proteins, or casein in offset formulations can cause a substantial increase in coating viscosity *(33)*. This effect is particularly profound with formaldehyde and formaldehyde-based products used at a high dose level and high temperature. Styrene-maleic anhydride resins, which are sometimes used with starch, can also cause significant increase in coating viscosity. Thus, some precaution is warranted before attempting the use of these types of insolubilizers.

Literature Cited

1. Engström, G. and Rigdahl, M., *Nordic Pulp Paper Res. J.* 1: 25(1989).
2. Young, T. S. and Fu, E., *Tappi J.* 74(4): 197(1991).
3. Whalen-Shaw, M., TAPPI 1989 Coating Conference Proceedings, TAPPI PRESS, Atlanta.
4. Clark, N. O., Windle, W., and Beazley, K.M., *Tappi* 52(11): 2191(1969).
5. Eklund, D.E. and Salminen, P., *Tappi J.* 69(9): 116(1986).
6. Sandas, S. E., Salminen, P. J., and Eklund, D., *TAPPI 1989 Coating Conference Proceedings,* TAPPI PRESS, Atlanta.

7. Baumeister, M. and Kraft, K., *Tappi* 64(1): 85(1981).
8. Engström, G. and Rigdahl, M., *TAPPI 1985 Coating Conference Proceedings,* TAPPI PRESS, Atlanta.
9. Fujiwara, H., Fujisaki, N., Shimizu, I., and Kano, I., *Tappi J.* 72(5): 91(1989).
10. Whalen-Shaw, M., *Tappi J.* 67(5): 60(1984).
11. Järnström, L., Ström, G., and Stenius, P., *TAPPI 1987 Coating Conference Proceedings,* TAPPI PRESS, Atlanta.
12. Napper, D. H., *Polymeric Stabilization of Colloidal Dispersions,* Academic Press, London, 1983.
13. Davis, R. M., *Tappi J.* 70(5): 99(1987).
14. Barber, E. J., *Tappi* 56(1): 52(1973).
15. Canard, P., *Tappi* 57(11): 95(1974).
16. Stinchfield, J. C., Clift, R. A., and Thomas, J. J., *Tappi* 41(2): 77(1958).
17. Herbet, A. J., *Review Methods to Measure Coating Immobilization Solids and Associated Coating Dehydration rates,* TAPPI PRESS, Atlanta, 1988.
18. Beck, U., Goossens, J.W.S., Rahlwes, D., and Wallpott, G., *TAPPI 1983 Coating Conference Proceedings,* TAPPI PRESS, Atlanta.
19. Taylor, D. L. and Dill, D. R., *Tappi* 50(11): 536(1967).
20. Herbet, A. J., Gautam, N., and Whalen-Shaw, M. J., *TAPPI 1990 Coating Conference Proceedings,* TAPPI PRESS, Atlanta.
21. Soemers, N. H., *Tappi* 53(4): 640(1970).
22. Jones, S. W. and Hetherington, L. E., *Tappi* 63(4): 137(1980).
23. Lonkila, S., thesis, Oulu University, Finland (1989).
24. Wilson, J. W., and Applegate, P. D., *Paper Trade J.,* June (1948).
25. Eklund, D. E., *TAPPI 1988 Coating Conference Proceedings,* TAPPI PRESS, Atlanta.
26. Tsuji, A., Nitta, J., Sasagawa, Y., and Nojima, N., *TAPPI 1990 Coating Conference Proceedings,* TAPPI PRESS, Atlanta.
27. Rahlwes, D., *Wochenbl. für Papierfabr.* 3: 83(1986).
28. Ramthun, J., Reif, L., Wallpott, G., and Rahlwes, D., *TAPPI 1989 Coating Conference Proceedings,* TAPPI PRESS, Atlanta.
29. Chapman, V. J., *Seaweeds and Their Uses,* 2nd edn., Methuen & Co., London, 1970.
30. Haug, A., Larson, B., and Smidsrod, O., *Acta. Chem. Scand.* 21: 691(1967).
31. Haug, A., Myklestad, S., Larson, B., and Smidsrod, O., *Acta. Chem. Scand.* 21: 768(1967).
32. *Hydrophilic Derivatives of Alginic Acid for Scientific Water Control,* Kelco Technical Handbook, 2nd. edn.
33. Windover, F. E. and Kelly, H. C., in *TAPPI Monograph Series No. 25,* "Coatings Additives," Chapter 9, 1963.
34. Funke, W. E., *J. Coating Tech.* 60, 767: 69(1988).
35. Greenwald, H. L., and Luskin, L. S., in *Handbook of Water-Soluble Gums and Resins* (R. L. Davidson, Ed.), McGraw Hill, New York, 1980, Chap. 17.
36. Morawetz, H., *Macromolecules in Solution,* 2nd edn., R.E. Krieger Publishing, Malabar, FL, 1983, Chap. VII.
37. Glass, J. E., *TAPPI 1988 Coating Conference Proceedings,* TAPPI PRESS, Atlanta.
38. Lundberg, D. J., Alahapperuma, K., Fernando, R. H., and Glass, J. E., *TAPPI 1990 Coating Conference Proceedings,* TAPPI PRESS, Atlanta.

CHAPTER 5: COLORATION OF AQUEOUS PIGMENTED COATINGS

Terrence L. McComb and Ralph H. Strickler

Introduction

Various types of colorants are used in water-based paper coatings to produce a wide range of colored papers ranging from tinted white to deeply colored specialty grades. The choice of suitable colorants will depend upon several factors. Some of the factors to consider are:

- End-use requirements such as brightness, various light and chemical fastness properties, opacity, etc.
- Compatibility of the colorants with the coating slurry.
- Ease of handling/working properties of the colorants.

The Science of Color

Color is caused by a combination of absorbance and reflectance of different parts of the visible light spectrum. Normal incident light can be thought of as consisting of all the wavelengths of light — all colors of light mixed together. When light strikes a white sheet of paper, all parts of this spectrum are reflected back and we see the sheet as visually white (in reality, no such perfect white exists). In a colored grade, certain wavelengths are absorbed and we only see the wavelengths that are reflected. A paper that is blue to our eye means that almost all incident light except the blue wavelength has been absorbed by the colorants. A red paper appears red due to almost all colors being absorbed except the red. We get different shades of blues and reds depending on how much of this visible spectrum is actually absorbed. By the same method, all other colored shades are achieved.

The mixing of colorants to achieve a desired shade is different from mixing wavelengths of light. As discussed above, if all wavelengths of light are present, we have white light. Conversely, if all shades of colorants are mixed together, we achieve a shade that looks black (absorbs all light). In coloring paper, or any other substrate, we achieve our visual result by subtracting various wavelengths of light. An example would be producing a green shade coating by the mixing together of a yellow and blue colorant. The blue colorant subtracts (absorbs) the red, orange, and yellow parts of the visible light spectrum. The yellow colorant subtracts the blue and violet parts of the spectrum and only green light is left to be reflected back to our eye.

In comparing reflectance curves in the visible spectrum (400–700 nm.) of a colored versus uncolored sheet, it can be seen that whenever a colorant is added, the area under this curve is decreased. For this reason, coloration is referred to as a subtractive process. Each specific colorant has its own unique reflectance/absorbance curve (**Figs. 5.1–5.6**), and a combination of colorants will give a reflectance curve with multiple absorbance dips indicating the colorants used (**Fig. 5.7**). It is this feature that is used in computer color matching.

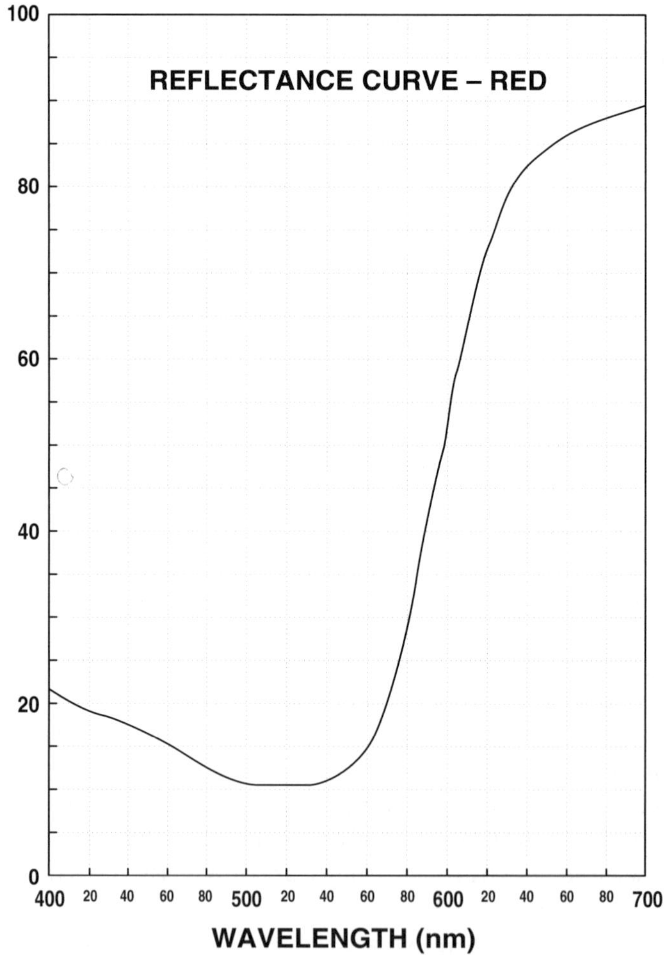

Fig. 5.1. The reflectance curve of a red colorant

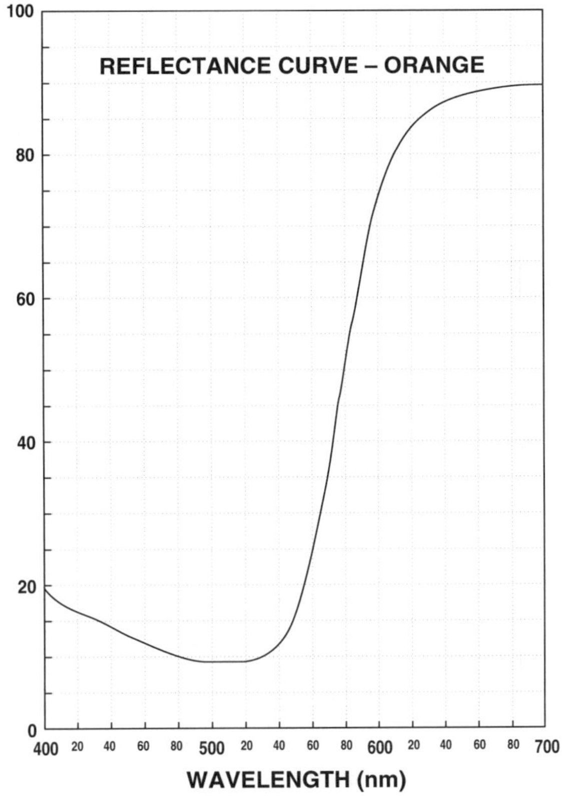

Fig. 5.2. The reflectance curve of an orange colorant

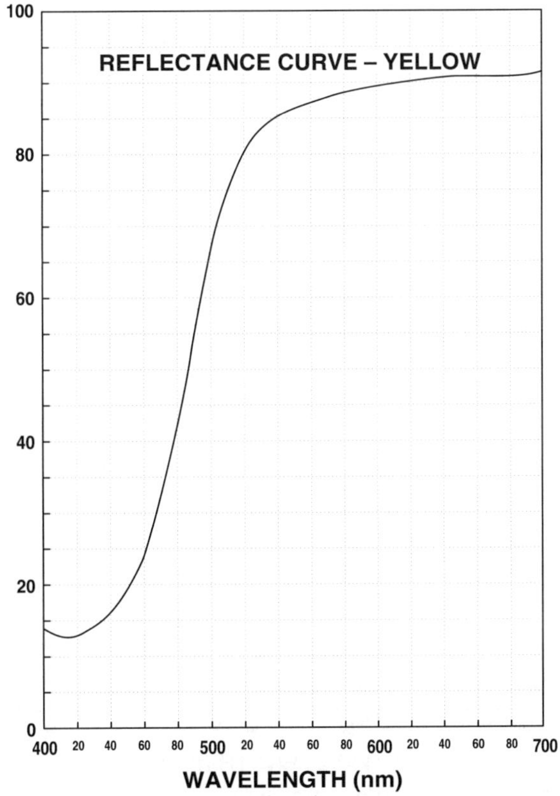

Fig. 5.3. The reflectance curve of a yellow colorant

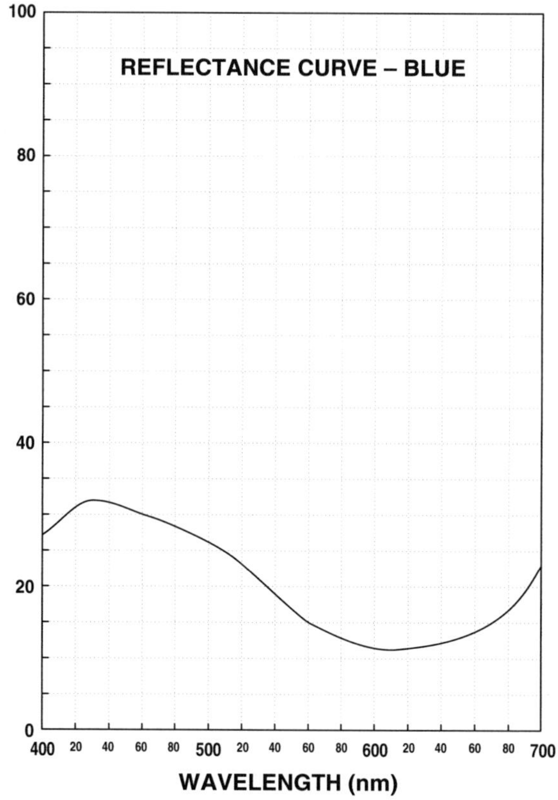

Fig. 5.4. The reflectance curve of a blue colorant

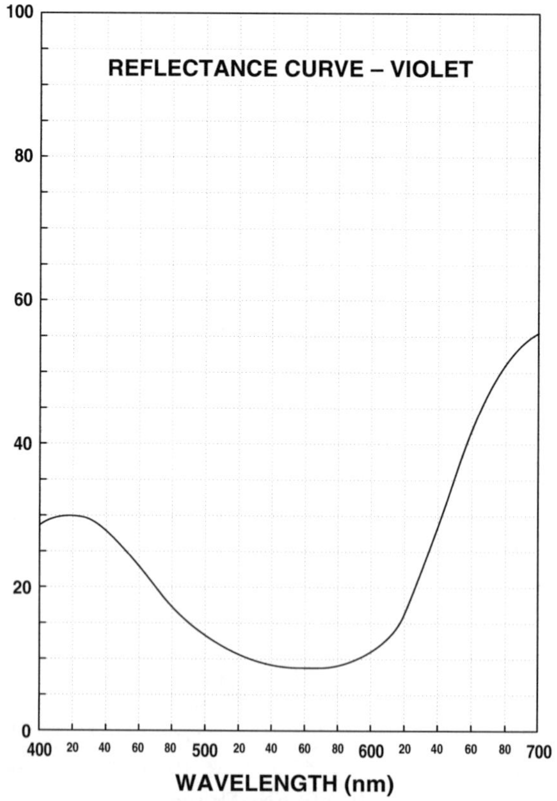

Fig. 5.5. The reflectance curve of a violet colorant

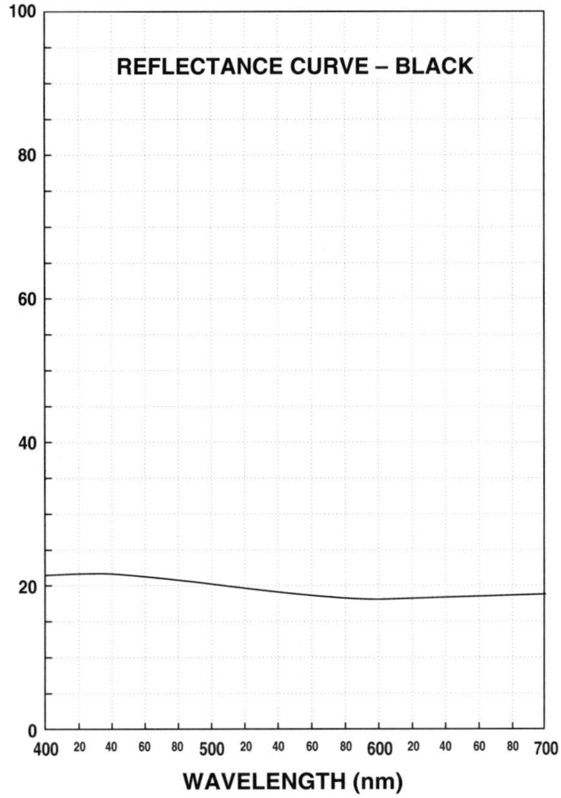

Fig. 5.6. The reflectance curve of a black colorant

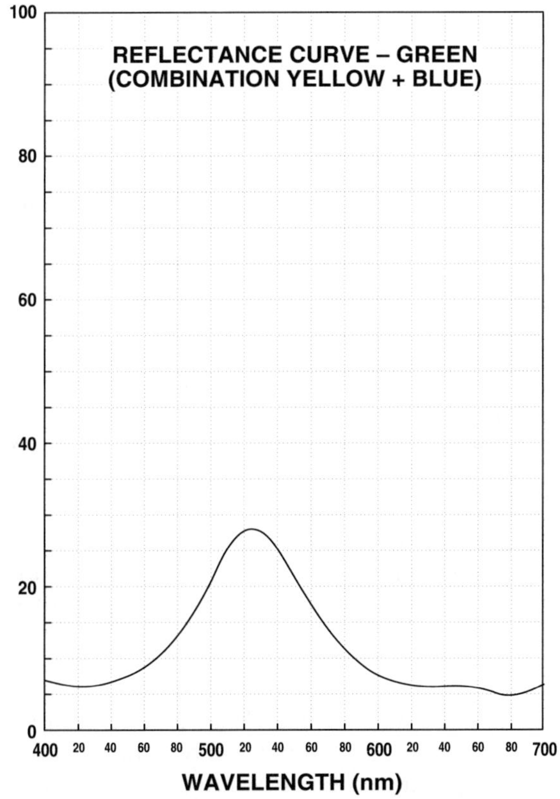

Fig. 5.7. The reflectance curve of a green produced with a combination of yellow and blue colorants

In contrast to this subtractive method of coloration using dyes and colored pigments, the use of fluorescent whitening agents (FWA) is an additive process of coloration. Fluorescent whitening agents are often added to white coatings, and often the basestock as well, to increase whiteness, brightness, and blueness, and give the finished coated sheet a visual "snap." FWA's work by absorbing the invisible ultraviolet light found in daylight and fluorescent light and emitting this absorbed energy in the blue region of the visible spectrum. When investigated by means of a reflectance curve, it is seen that the area under the curve is increased compared to a sheet with no FWA. The peak area of this increased reflectance also coincides closely with the wavelength at which TAPPI brightness is measured (457 nm). While blue, violet, and red pigments and/or dyes are often used in the tinting of white shades to achieve a "whiter" look, these colorants actually decrease brightness due to their subtraction (absorption) of light energy. The mechanism of the fluorescent whiteners actually does increase brightness as it adds to reflectance rather than subtracts from it. In actuality, producers of white papers will often use a combination of both FWA and colorants to be able to increase brightness and attain and control a desired shade.

The Terminology of Color

Color can be described scientifically with the use of several terms and various color spaces. The most commonly used system is instrumental evaluation using an "L, a, b" system. "L" represents strength (white to black scale). The "a" reading describes redness or greenness of shade, while the "b" reading represents blueness or yellowness. Readings on a production sample are compared versus the shade standard, and the colorants are adjusted to achieve a degree of color difference (Delta L, a, b) which is within an established tolerance of acceptability. Often a "Delta E" reading (total color difference derived from the L, a, b readings) is also used for determining acceptability.

In comparing colored samples of coated grades, a sample and standard are compared versus one another using the more common terminology of strength, shade, and brilliance.

Strength

Strength is the designation of depth of shade, and is used as a percentage measurement. For example, a particular red shade might be 20% stronger than one sample and 10% weaker than another. For strength evaluations to be accurate and meaningful, the colors being compared must be close in shade. It becomes very difficult when trying to compare the strength of a yellow versus the strength of a blue, and only general comparisons can be made in a case like this.

Shade

The science of color measurement does not use the many descriptive terms often applied to finished products and mill shade lines, such as mauve, cherry, lime, etc. All shades are compared on the basis of the three primary colors (red, yellow, blue). For example, a blue cannot be bluer — this would indicate a strength comparison rather than a shade comparison. A blue can be redder or yellower (greener). A red can be bluer or yellower. A yellow can be redder or bluer (greener). A green can be bluer or yellower. White, as it is a reflectance of all colors, can be redder, bluer, or yellower. Black and brown, as they exhibit varying degrees of absorbance of all colors, can be described by the same parameters used for white.

Brilliance

Brilliance is a term used to differentiate between brightness and dullness. While visual comparisons of brightness are often relative and arbitrary, color measurement instrumentation allows us to put a numerical value on these measurements. Instruments such as color difference meters, brightness meters, and other types of spectrophotometers allow us to put a value on not only brightness but color as well.

Coating Basestock

The results obtained by the application of a colored coating are influenced not only by the optical and physical properties of the base sheet, but by the base sheet appearance as well.

Except in the case of heavyweight coatings, mineral coatings will mask but not totally cover the brightness and hue of the base sheet. High spots, varying absorbency across the sheet, nonuniform density, etc. of the base sheet will affect the coating coverage and the extent of the base sheet "show-through." In tints and light shades, it is therefore important that the basestock have a brightness and/or hue as close as possible to that of the coating. In many cases, it may be necessary to color or brighten the basestock to be similar to the coating to avoid the problem of show-through and color variations (mottling) in the finished sheet. Even the application of a blue-white coating over a cream-white base may cause a problem due to a nonuniform coating application.

Classes of Colored Pigments for Coatings

General Information

The principle colorants used in aqueous coatings are the dispersible colored pigments, both synthetic organic and inorganic types. Although low in cost, the natural earth pigments, such as yellow ocher, burnt umber, and red iron oxide are often avoided due to their inherent lack of shade reproducibility.

In contrast to water-soluble dyes, colored pigments are composed of finely ground solid particles and are available as highly dispersible liquid, paste, and powder forms. Dispersible pigment colors may be diluted with water prior to use or added directly to and dispersed in the coating mixture. Due to their nondusting nature, the trend has been toward the use of the predispersed types.

The particle size of colored pigments is critical since the smaller the particle, the greater the available surface area of colorant, and the greater the tinctorial strength (color strength) of a given weight of pigment. An example of this would be comparing the surface area that could be covered with a pound of pebbles versus that which could be covered with a pound of sand. Large particle size, insufficient dispersion, or agglomeration (clumping together) of colored pigment particles can cause color streaks and specking when these large particles are drawn across the sheet during the coating operation. A similar effect can result from incompatibilities between the colored pigment dispersion and the coating mix. While the majority of predispersed colored pigment pastes and liquids in use are anionic dispersions, there are some forms available which include an amphoteric dispersant to help alleviate problems with incompatibility.

As a class, colored pigments are of lower tinctorial strength than the water-soluble dyes, but generally exhibit a higher degree of lightfastness, bleedfastness, and are more resistant to the effects of chemicals such as acids, alkalis, and chlorine.

Major differences between organic and inorganic pigment colorants can be summarized as follows:

1. Organic types are generally more expensive, but often compensate for this by being brighter and of higher tinctorial strength.
2. Inorganic types generally exhibit better hiding power (more opaque), better lightfastness, and better chemical stability.

Inorganic Colored Pigments

The inorganic pigment colorants are based on the natural minerals mentioned previously, processed minerals, or synthetic products. The synthetics are generally preferred due to their uniform composition and particle shape, narrow particle size distribution, and easy dispersibility in water-based systems. As stated above, the advantages of these colorants are low cost, high lightfastness, and excellent hiding power. Some of the more commonly used inorganic pigment colorants are:

- Chrome Yellow (Pigment Yellow 34)
- Iron Oxides
- Molybdate Orange (Pigment Red 104)
- Cadmium Red (Pigment Red 108).

Included in this category are the phosphorescent pigments, such as those based on zinc and cadmium sulfide, and also those based on silicates and titanium compounds which give transparent, pearlescent, or iridescent properties.

Organic Colored Pigments

The organic colored pigments are considered most important for use in water-based pigmented coatings. Predominantly used are the azo, anthraquinone, and phthalocyanine types. Some of the more important colored pigments are listed by category as follows:

1. Azo Pigments

- **Hansa Yellows (Pigment Yellows 1, 3, 4, 6).** These are bright yellows with high tinctorial strength. Although they exhibit relatively low hiding power (opacity), they are usually considered to be the most opaque of the organic yellows. They exhibit good resistance to acids and alkalis.
- **Diarylide Yellows (Pigment Yellows 12, 13, 14, 17).** Generally higher tinctorial value than the Hansa Yellows but with inferior lightfastness to the Hansa types.
- **Hansa Orange (Pigment Orange 1).** A high strength color with good lightfastness in deep shades.
- **Dinitraniline Orange (Pigment Orange 5).** Tinctorially strong, superior in this respect to the inorganic molybdate oranges.
- **Pyrazolone Orange (Pigment Orange 13).** A color with good chemical fastness but low in hiding power.
- **Para Red (Pigment Red 1).** Low cost, high tinctorial value, good hiding power and good alkali resistance. Solvent bleedfastness is generally considered poor.
- **Toluidine Red (Pigment Red 3).** Good alkali fastness and hiding power, poor bleedfastness in many solvents.
- **Chlorinated Para Red (Pigment Red 4).** Sometimes referred to as “Fire” or “Flaming” red. Superior lightfastness but only average hiding power.
- **Pyrazolone Red (Pigment Red 38).** A disazo pigment with high tinctorial strength.
- **Lithol Reds.** These are actually a family of precipitated azo pigment colors. Used widely where high tinctorial value, high hiding power, and low cost are important. These generally have poor alkali stability.

- **Bon Reds and Maroons (Pigment Red 48, 52, 57).** Based on B-oxynapthoic acid (Bon) — a range of bright colors with good fastness properties.

2. Phthalocyanine Pigments

- **Copper Phthalocyanine Blue (Pigment Blue 15).** Excellent color value, excellent lightfastness, chemical and heat stability. Low in cost.
- **Copper Polychlorophthalocyanine Green (Pigment Green 7).** Possesses the same excellent characteristics as the blue.

3. Thioindigo Pigments

These vat pigments range in shade from a yellow shade red to violet. Colorants in this group vary considerably in lightfastness, bleedfastness, and heat stability.

4. Anthraquinone Pigments

Some pigments from this group have gained acceptance as colorants for paper coatings:

- **Indanthrone Blue (Vat Blue 4/Pigment Blue 60).** Redder than the phthalocyanine blues and exhibits good fastness properties.
- **Isodibenzathrone Violet (Vat Violet 1).** Blue-shade violet with high tinting strength and excellent chemical resistance comparable to the copper phthalocyanine pigments. Often used in paper coatings as a tinting colorant for white grades, exhibiting good lightfastness properties.
- **Anthanthrone Orange (Vat Orange 3, Pigment Red 168).** Offer excellent lightfastness and chemical resistance but are high in cost.

5. Miscellaneous Pigments

- **Pigment Green B (Pigment Green 8).** A complex iron salt with high tinctorial strength and low cost.
- **Carbon Blacks.** It is debatable whether or not these should be termed inorganic or organic. As these are generally produced by combustion of organic materials, they are here listed with the organics. Carbon blacks are widely used due to their low cost, high lightfastness, high opacity, and chemical stability. They are often used as dulling agents due to their neutral shade, dulling while not greatly affecting shade.
- **Carbazol Violet (Pigment Violet 23).** High tinctorial value and high lightfastness. Good colorant for both tinting and in deep colors. Often used as a shading component in blue shades.

Lakes

Lakes are organic colorants derived from water-soluble acid, basic, or direct dyes, made insoluble by treatment with suitable inorganic or organic compounds. With respect to paper coatings, the most important of these types are the lakes resulting from treatment of basic dyes with complex inorganic acids, such as phosphomolybdic, phosphotungstic, or phosphotungstomolybdic acids.

The resulting colorants are brilliant, relatively strong, and have improved, though still only fair light, heat, and water bleed resistance compared to the basic dyes from which they are derived. Generic names of some of the more important products are:

- Pigment Red 81
- Pigment Violet 3
- Pigment Blues 1,2,14
- Pigment Green 1.

The reaction with the inorganic acid used to produce these colorants is reversible, so care must be taken in their use. Conditions of high pH (8.5 and higher) and elevated temperature may cause the original basic dye to be reformed, resulting in strength and shade variations on the coater and poorer lightfastness. Lakes of basic dyes are generally not recommended for latex or acrylic coatings, where high pH conditions are necessary.

Soluble Dyes and Fluorescent Whitening Agents

General Information

While the dispersible colored pigments discussed above consist of generally insoluble colored particles, the dyestuffs used in coatings are water soluble. While the colored pigments are thought of as having no affinity for coating minerals such as clay or calcium carbonate, the water-soluble dyes exhibit varying degrees of affinity for these minerals and will actually bond to the minerals. In the case of TiO_2, this is generally considered inert and, with some exceptions, will not be colored by the water-soluble dyes commonly used in paper coatings.

Economically, the use of dyes as coating colorants is much more advantageous than the use of colored pigments. Dyes as a whole offer much higher color strength than the colored pigments and, although dyes generally do not offer the high fastness properties or opacity found with the colored pigments, their economy merits their use in grades where these properties are not as critical to the end use of the coated paper. Water-soluble dyes are bleachable in varying degrees, so that the use of dyes also affords the possibility of bleaching colored broke — normally not achievable with colored pigments.

Acid Dyes

The acid dyes are usually sodium, potassium, or ammonium salts of the corresponding color acids, such as the sodium salt of a sulfonic acid derivative. The acid dye chromophores are anionic in charge, and as a class these dyes generally exhibit very high solubility and brightness of shade. Most of them have only fair to good lightfastness. Their high solubilities result in limited use in the coatings area due to their tendency to remain with the aqueous phase of the coating mix and resultant poor to fair bleedfastness. This high solubility has one benefit — they can be used to produce shades of greater depth than the lower solubility direct dyes. However, the bleedfastness may be so poor that even a clear coat may provide only fair water bleedfastness.

Basic Dyes

Basic dyes are typically salts, such as chlorides, sulfates, hydrochlorides, and oxalates of a color base. Basic dye chromophores are cationic in charge and are sensitive to free alkali. As a class, the basic dyes are probably the brightest, lowest cost colorants available, but they have many drawbacks which severely limit their use in coatings. They possess only fair solubility in water and exhibit poor light and bleedfastness properties. If used at all, they are normally used in light shades only. At higher levels, their cationic charge tends to cause flocculation of the coating mix due to incompatibilities with casein, protein, latex, and anionic dispersants. At lower concentrations, they may show a tendency to lower viscosities of slurries containing casein or protein. Many are found in liquid form, which may increase the chances of flocculation or incompatibility due to these liquid forms using acids to keep the dyes in concentrated solution.

Anionic Direct Dyes

The direct dyes are the salts of dye acids. They are similar to acid dyes in being anionic in charge, but the molecules are more planar in shape, and solubilities are generally lower than the acid dye class. They are known as direct dyes since they will, in most cases, readily exhaust on cellulose; and, due to this, they are the most widely used dyes for the wet-end dyeing of fine papers made from bleached fiber. The direct dyes are next in importance to the colored pigments for the coloration of aqueous pigmented coatings. Although generally not as bright as the acid or basic dye classes, they exhibit higher bleedfastness properties compared to these classes. Lightfastness varies within the class, but as a class they exhibit higher lightfastness than the acids and basics, and lower lightfastness compared to the colored pigments. Due to their greater affinity for clays, they exhibit much less tendency than the acid dyes to bleed or remain in the aqueous phase of the coating mix. Being anionic in charge, they do not give the compatibility problems that are present with the basic dyes.

Direct dyes are available in both dry and aqueous liquid forms. As with pigments, the trend is toward the liquid forms for ease of handling and to avoid problems with solubility. They can be used alone or in combination with the colored pigments or acid dyes. They are normally compatible with protein, casein, and latex binders and are stable under alkaline conditions. In coatings containing polyvinyl alcohol as a binder, care should be taken to use direct dyes which do not contain borax as a diluent, to ensure compatibility with the binder.

Cationic Direct Dyes

This is a relative newcomer to the dyestuff classes, and has not yet been officially designated as a dye class. Developed mainly for the wet-end dyeing of paper fiber, this dye class is generally similar to the direct dye class, except that these dyes have a cationic chromophore in place of the anionic chromophore found on direct dye molecules.

Being cationic, the cationic direct dyes do pose some of the compatibility problems found with the basic dyes, but offer advantages which may merit their increased use in paper coatings in the future. Compared to the older technology basic dyes, the cationic direct dyes offer very good water and alcohol fastness, and superior lightfastness. Generally, the cationic directs do not have quite as high lightfastness as the anionic direct dyes. As with the direct dyes, they are available in dry and liquid forms.

To avoid compatibility problems, the cationic direct dyes must be added to the coating mixture (makeup water plus a partial white pigment addition) and allowed to chemically bond with the white pigment before any dispersants are added. They exhibit very quick exhaustion and, shortly after their addition, the dispersants and additional white pigments and other chemicals may be added.

Fluorescent Whitening Agents

As discussed previously, the fluorescent whitening agent (FWA) is used to actually increase whiteness and brightness by its absorption of invisible ultraviolet (U.V.) light and emission of this absorbed energy in the blue region of the visible spectrum (**Fig. 5.8**).

Although all FWAs emit visible light in the blue region, there can be subtle shade differences between different products depending on the specific point at which this light is emitted. Shades range from greenish blues to those with a reddish cast.

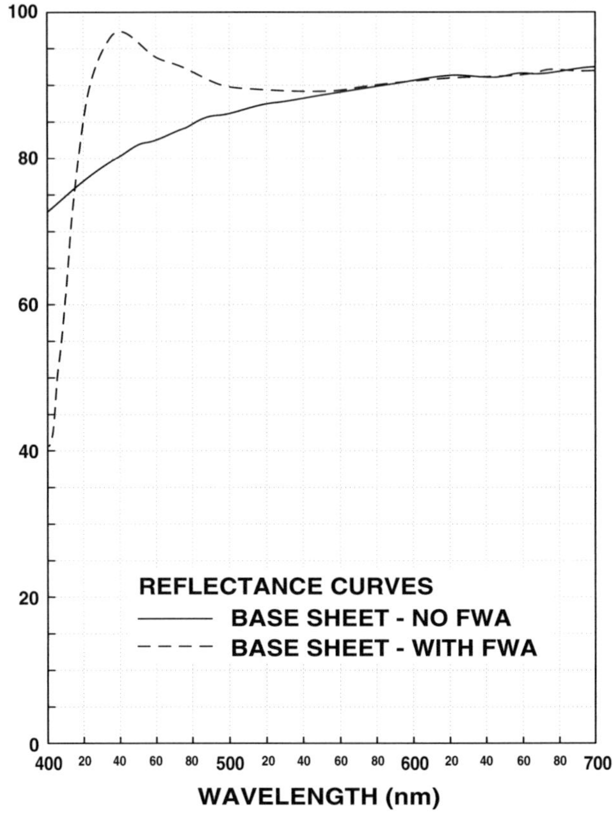

Fig. 5.8. Reflectance curves illustrating the emission of light in the blue region by a fluorescent whitening agent

The fluorescent whiteners used in aqueous coatings, and in wet-end and surface applications of base sheets, are anionic in charge and are of stilbene-based chemistry. They are very similar in molecular structure to the anionic direct dyes and are usually available as aqueous liquids. Lightfastness of the water-soluble FWAs commonly used by the paper industry is generally considered only fair.

Be aware that, while titanium dioxide is generally considered the brightest opacifying agent available, when used in conjunction with TiO_2, its use adversely affects FWA efficiency. Titanium dioxide is a strong absorber of ultraviolet light energy, and competes with the FWA for this energy, absorbing energy that the FWA needs for optimum efficiency. Of the two types of TiO_2 available, anatase and rutile, the anatase is preferred for use with fluorescent whiteners. While both types of TiO_2 are strong U.V. absorbers, the rutile form also absorbs more visible light, lowering visual reflectance more than the anatase form (**Fig. 5.9**). While the use of TiO_2 and FWA are often necessary to achieve a desired result, this fact should be kept in mind, as levels of FWA may need to be increased substantially when TiO_2 is introduced or increased in the coating mix, or when rutile TiO_2 is used in place of anatase. A more economical approach to using high amounts of TiO_2, if possible, would be to increase opacity with a higher percentage of clay and/or aluminum trihydrate, and use a FWA to get desired brightness. While clay does absorb some U.V. energy, it is a much lower absorber than titanium dioxide.

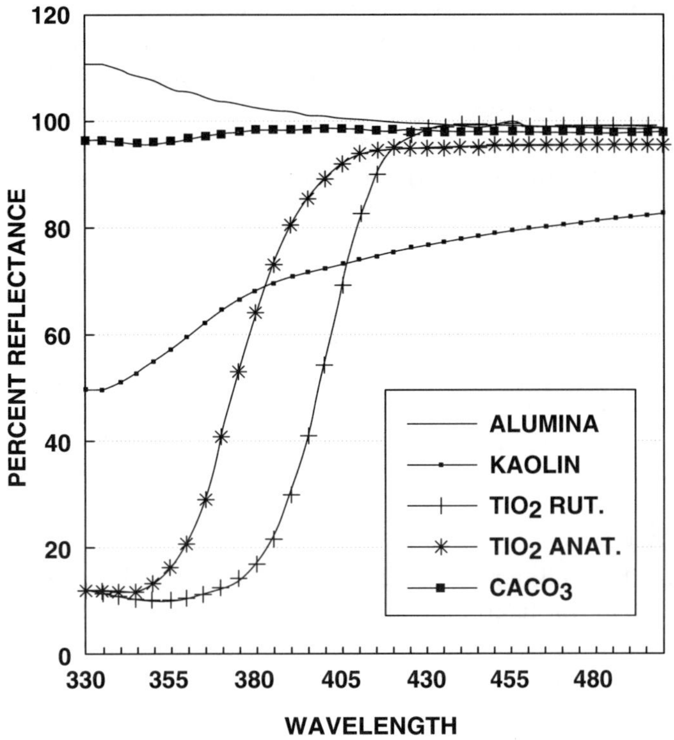

Fig. 5.9. Reflectance curves of titanium dioxides vs. other minerals. Curves illustrate absorbance in U.V. region below 380 nm.

As with water soluble dyes, there is a saturation point for the fluorescent whiteners beyond which additional brightness cannot be obtained. Adding additional FWA beyond this point will result in a yellowing/greening of shade and may result in loss of brightness. This greening over is due mainly to fluorescence quenching by UV-absorbing impurities in the FWA which can become evident at high levels. The amber color of the FWA actually has less to do with this quenching effect than previously thought. Low pH of the coating, although rare, could cause similar problems.

Several products are available for use in coatings. Some are designed to work more efficiently with synthetic binders; others are more applicable in producing the super high brightness whites. Whatever the application, suppliers of these products should be contacted as to which product or products are best for your application.

Addition of Colorant

With the exception of the basic and cationic direct dyes, most dyes, fluorescent whiteners, and colored pigments may be added to the coating mix at any stage. It is essential that adequate mixing follow their addition to provide uniformity of coating mix coloration. It is often convenient to add the colorant or FWA to the pigment slurry before dispersion in a mill. The highly dispersed pigment colors should not undergo any shade change in this operation because their degree of dispersion is not increased during further milling. Milling of a pigment slurry containing the soluble dyes should likewise have no adverse effect on shade.

Lightfastness

The lightfastness of tinted or colored coated papers is dependent, not only upon the lightfastness and concentration of the colorants, but also upon the base paper, coating pigments, and binder. Generally, among the colorants used in aqueous paper coatings, basic dyes have the poorest lightfastness, pigments the highest, with the other colorants being between these two classes.

Base papers for coatings made from groundwood or semibleached pulps will tend to revert rather rapidly, affecting the shade of the coating as well if show through is present. Therefore, colors with a high degree of lightfastness are probably not needed for these base papers. On the other hand, base papers made from bleached chemical pulps, having much less chance of reversion, would call for the improved lightfastness of the direct and pigment colors.

A slight fade after four to eight hours in the standard carbon-arc Fade-Ometer is usually considered adequate for most grades of coated paper. Color lightfastness is relative to the depth of the shade, with deeper shades exhibiting more resistance to fade than lighter shades. Special end uses demanding high fastness for extended periods will require special attention to the selection of base papers, binders, and colorants.

Color Matching

It is recommended that mill personnel rely on both visual and instrumental comparisons for accurate shade control. Instrumental measurements provide impartial evaluations of color shades and allow reporting them numerically. Visual judgment is still required in the mill as it is usually much more rapid than instrumental measurement (except where on-line color measurement instrumentation has been installed), and provides an essential check on the accuracy of instrument readings. Also, the ultimate end-use customer will probably be judging color on a visual basis.

Visual judgments, however, can be influenced by many factors. For example, different illuminations, surroundings (colored walls, etc.), and the mood and age of the observer will affect visual judgments. Two observers will often disagree on the shade differences between two colored samples, due to slight differences in color vision. These variations in human color vision can be avoided with instrumental measurements. Instrumental measurements are also of particular interest in establishing color specifications and tolerances on a quantitative, numerical basis.

Currently, the most common instruments for measuring color differences are those which express measurements as numerical Hunter Lab, CIE Lab, brightness, and whiteness readings. Instruments are also available which in addition to these common readings will generate reflectance curves of the colored sheet and other pertinent spectrophotometric data.

For visual judgments, it is necessary that a standard artificial daylight type lamp be used for illumination of the samples. The use of a standard light source eliminates the variations in judgment which are caused by changes in the light source, permits making reliable judgments at any time, and helps assure that shades from one mill will match shades from another.

It is a well known fact that if a shade is matched with two different colorant formulas, the possibility exists that the papers may appear identical under one light source but very different under another. This is known as *metamerism* and is referred to as a change of shade under "artificial" or "incandescent" light. A number of suitable light sources are available which contain different, switchable lamps to allow for checking for metamerism. These light sources usually contain a combination of daylight, horizon (incandescent), fluorescent, and ultraviolet lamps. They should be installed in an area protected from dirt and extraneous light. Often, it may be necessary to change color formulas due to cost, runnability problems, environmental changes, etc. In these cases, some degree of metamerism may be unavoidable, and new standards may have to be agreed upon by mill and customer, based on a visual examination under a predetermined light source. As instrumental readings will not always reveal this metamerism, new measurement reading standards and tolerances, based on a visually acceptable match, may have to be set up as well.

The process of color matching itself can be accomplished at the mill by trained personnel or it can be done by the technical support personnel of a colorant supplier. It is recommended that the process be first carried out on a lab scale to be sure that he combination of contemplated colorants can indeed be used to produce the desired shade, and to determine what amounts of each color will be needed. Trying to match a shade on the machine, guessing which colorants and amounts are needed, can be difficult and time consuming. As with any lab work, it is very important to duplicate mill conditions as closely as possible, in order to get a lab formula which will repeat closely on the paper machine. Concerning paper coatings, this means that the coating makeup and base paper used in the lab should be identical to that which will be used in the mill. This is easy to accomplish if the lab work is done at the mill, where the materials are readily available, but is a different story if done in a supplier's lab. As coating makeup formulas are usually very proprietary, the best solution for off-site lab work is to send a shade standard, base sheet paper, and uncolored slurry to the supplier's lab. This will eliminate the necessity for releasing proprietary information, while allowing the supplier to work on a coating mix which is identical to that which will be used in the mill, eliminating many of the lab to machine variables. This is especially critical for white grades and tints, where coating components are controlling factors in the resultant shade.

Bibliography

1. *Colour Index,* 2nd edn., Vol. 2, The Society of Dyers and Colourists, Bradford, England, and American Association of Textile Chemists and Colorists, Research Triangle Park, NC, 1982.
2. *Dyeing of Paper,* E.I. du Pont de Nemours, Wilmington, DE, 1969.
3. *Encyclopedia of Chemical Technology,* 2nd edn., Vol. 15, John Wiley and Sons, 1968.
4. Lubs, H. A., Ed., "The Chemistry of Synthetic Dyes and Pigments," American Chemical Society Monograph Series, Reinhold, New York, 1955.
5. "Paper Coating Pigments," Monograph No. 30, TAPPI PRESS, Atlanta, 1966.
6. "Paper coating additives," Monograph No. 25, TAPPI PRESS, Atlanta, 1978.

CHAPTER 6: DISPERSANTS

Ken L. Snyder

Introduction

Paper is coated with mineral particles to improve the optical properties and printability of the finished paper product. The mineral pigments used for this application include kaolin, calcium carbonate, titanium dioxide, and others. It is very desirable to have these minerals supplied as high solids aqueous dispersions. In their use, the dispersed pigments are mixed with other coating components and applied to the basesheet as a thin high solids film. Excess coating is often removed by a blade or air knife. During the process, the coating must be fluid enough to flow under the blade without scratching or streaking. In general, it is advantageous to coat paper at high solids. High solids coatings minimize drying requirements, limit the migration of materials with an affinity for water, and optimize coating structure.

Dispersants are chemicals used in the deflocculation process of coating pigments. These chemicals increase the surface potential of the individual mineral particles and thereby:

- Reduce the energy necessary to separate minerals, like kaolin, into discrete particles of high aspect ratios
- Provide stability that prevents agglomeration upon storage of the clay slurry
- Provide a minimum coating viscosity at high solids when used at the proper dosage level.

The various types of unique compounds that are classified as dispersants play subtle but significant roles in paper coating behavior. The effect of the dispersant on the rheology or flow properties of the pigment slip can be demonstrated by a simple experiment. In this experiment, 70 g of dry kaolinite clay is placed in a beaker, and 30 g of distilled water is stirred in to thoroughly wet out the clay particle surfaces. At this point, it will be observed that a very thick paste is obtained, so thick that the beaker can be inverted and none of the clay–water mixture will pour out. To this paste, a few milliliters of an aqueous solution of a polyphosphate dispersant are added, and, after gentle stirring, the suspension becomes sufficiently fluid so that it will readily pour from the beaker. This experiment can be repeated with each of the paper coating pigments, and, although the phenomenon is still not completely understood, it is of practical significance because the low viscosity that is obtained permits the coating color to be prepared at very high solids. This eliminates the need for, and resulting cost of, removing large volumes of water during drying. It also permits the coater to operate at high speed since drying capacity, following application of the coating, frequently limits coating speed.

The pigment dispersant, through its influence on solids content during coating, also affects the mode of consolidation of the drying film and, hence, its structure when dry. The dispersant thus affects the pore volume and pore size distribution of the dried coating film. This in turn influences the ability of the coating to receive ink during printing and to effectively scatter light. The efficiency of light scattering can be quantitatively related to the brightness and opacity of the coated sheet. From this discussion, it can be seen that although the dispersants are usually used in amounts of less than 1% based on the weight of the pigment used in the paper coating formulation, they play a most important role in the overall performance of the coating film.

Classification of Dispersants

While there have been many different chemical compounds suggested as dispersants in the literature, for this discussion we shall concern ourselves with only five broad classifications. They include:

1. The anionic polymer dispersants
2. The polyphosphates
3. The alkali silicates
4. The alkalies
5. The nonionic polymer dispersants.

We shall consider each of these classifications from their chemical structure, relative effectiveness as dispersants, handling considerations, and method of manufacture.

The Anionic Polymer Dispersants

Before going into detail on the anionic polymer dispersants, it is important to draw some important distinctions between dispersants and surfactants:

Differences Between Dispersants and Surfactants

Dispersants	Surfactants
*Little effect on surface tension	*Greatly reduce surface tension
*Low foaming	*Foamy
*Little wetting action	*Powerful wetting
*High deflocculating power	*Little deflocculating power
*Polymeric with numerous ionic groups	*Low molecular with one ionic group

Surfactants are low molecular weight compounds consisting of a hydrophobic portion (tail) and an anionic end group (head). Comparatively, anionic polymers are high molecular weight materials in which many anionic groups have been polymerized at random along the polymer backbone. This distinction is made due to the differences the two types of chemicals have on wetting, surface tension reduction, and ability to cause foam.

The anionic polymer dispersants are by far the most commonly used class of dispersants. They have gained in popularity in recent years due to their stability in clay slurry storage for long periods of time. Some other classes of dispersants suffer hydrolysis and require the use of additional dispersant or auxiliary dispersant. In today's paper coating mills, product quality and consistency are extremely important. Again, since clay slurries prepared with anionic polymer dispersants do not change in viscosity with time (see **Fig. 6.1**), they have become the dispersant class of choice both in the clay mines and the coating kitchens.

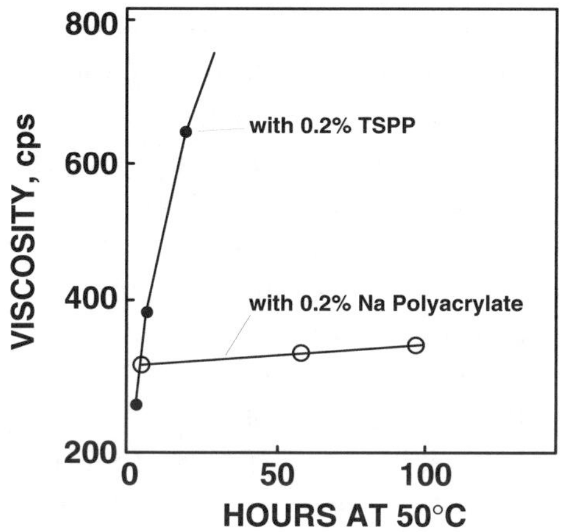

Fig. 6.1. Viscosity stability of clay/latex blends

The most common type of anionic polymer dispersants are the sodium polyacrylates. This family of polymers is produced by the copolymerization of carboxylic acids (i.e., acrylic acid) with acrylic acid esters (i.e., ethylacrylate, butylacrylate). Changing the ratio of these two species adjusts the amount of anionic charge per gram of polymer. Another very important property of such polymers is their degree of polymerization or molecular weight. **Figure 6.2** depicts the mechanism of vinyl polymerization.

Initiation:	**Thermal Initiator**			→	**Free Radicals (R)**
Propogation:	**R**	+	M_I	→	RM_I
	RM_N	+	M_I	→	$RM_{N \cdot I}$
Termination:	RM_X	+	RM_Y	→	$RM_{(X \cdot Y)}$ **(Combination)**
	RM_N	+	**CTA**	→	RM_N**CAP (Transfer)**

Where: M_I = **monomer molecular**
M_N = **n monomers copolymerized**
n = **degree of polymerization**

Fig. 6.2. Mechanism of vinyl polymerization

The above depicts the general mechanism of vinyl polymerization. Monomer molecules used can be carboxylic acids (i.e., acrylic acid) or other vinyl monomers. Molecular weight is controlled by the amount of initiator and chain transfer agent (CTA) used in the polymerization process.

Molecular weight and composition are two of the most important polymer properties in determining its effectiveness as a dispersant. To fully understand why these properties are important, let us first consider the mechanism by which such ionic dispersants function.

Mechanism of Stabilization

The following is the mechanism by which ionic dispersants function. This theory is based upon a change in the magnitude of electrostatic charge on the particle surfaces as a results of preferential ion adsorption.

With very few exceptions, the common dispersants ionize when dissolved in water. The dispersants usually used in paper coating form large anions that may be strongly adsorbed on the surface of the pigment particles. The theories that best explain the phenomena of dispersion by ionic dispersants are those derived for colloids. The dimensions of pigment particles used in coating colors range from a few hundredths of a micron, for the crystals that compose the particles of titanium dioxide, to several microns for the particles of coarse clays, or ground calcium carbonate. The particles of the suspensions on which the classical experiments were made, which led to the theories accounting for the stability of these suspensions, were of the order of perhaps 0.01–0.1. There is no evidence that the phenomena of dispersion of pigments are changed by particle size alone. The pertinent theories can be found in the literature under, for example, *Theory of the Stability of Lyophobic Colloids (1)*. In all the considerations of this discussion, the subject is restricted to the stability of hydrophobic colloids.

The common inorganic pigments used in paper coating are hydrophobic because they are not significantly swollen by water and their crystalline particles maintain their individual size and shape. This is to distinguish them from the class of hydrophilic colloids that have a strong affinity for water and are highly swollen by it. Starch, casein, and soya-bean protein are important examples of hydrophilic colloids used in coating colors.

The action of a dispersant begins when its ions are preferentially adsorbed from solution onto the surface of the pigment particles. As a result of this adsorption, the particle receives an electrostatic charge equal to the charge of the adsorbed ion, and it is localized at the surface of the particle. The magnitude of the charge depends upon the number of ions adsorbed and upon the charge of each ion. The dispersant in solution furnishes both positively and negatively charged ions, and the adsorption, which gives the particle an electrostatic charge, will remove a number of ions of one charge from the solution and leave an equal number of ions of opposite charge in the solution. There is not change in the net balance of positive and negative charges in the suspension or solution as a whole, but the action of the dispersant causes a redistribution of the charges. The pigment particles have an extra concentration of ions of one charge on their surface, and the solution surrounding the particle, next to the surface layer, has an extra concentration of ions of the opposite charge. The ions on the pigments surface attract the oppositely charged ions that are in excess in the solution, and they tend to cluster around the particle surface. Thus is formed the ionic electric double layer that is the basic concept of the classical theory of the stability of lyophobic colloids. A schematic diagram of the electric double layer on a pigment particle surface is shown in **Fig. 6.3**.

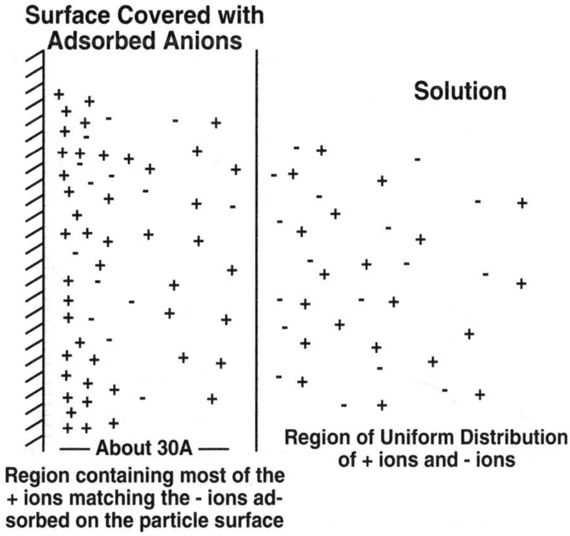

Fig. 6.3. Distribution of ions in the electric double layer

The ions that are in excess in solution, as a result of the adsorption of oppositely charged ions on the pigment surface, are called counter-ions. The counter-ions will not attach themselves to the oppositely charged ions on the surface, in spite of the strong electrostatic attraction. This is because the factors that caused the dispersant to ionize in the first place make some separation of positively and negatively charged ions a stable condition.

It is the counter-ions that are the direct cause of the repulsion between the particles of a suspended pigment stabilized by a dispersant. The creation of this repulsion is the property that makes a substance a dispersant. The counter-ions distribute themselves in a layer around the pigment and are held there by the attraction of the adsorbed ions. The important property of this layer of counter-ions is that it is quite thick. The counter-ions do not approach one another closely, nor do they cluster tightly against the adsorbed ions on the pigment surface. Rather, they distribute themselves outwardly from the pigment surface, with the greatest concentration next to the pigment surface; their concentration diminishes as the distance from the surface increases. The adsorbed ions on the pigment surface have concentrated their electrostatic charges in a layer only one ion thick, i.e., a few Ångstroms thick. The counter-ions have their charges dispersed through a layer that is some tens of Ångstroms thick.

When particles of pigment approach one another, as though to flocculate, the diffuse layer of counter-ions accompanying one particle begins to merge with the corresponding layer of the other particle. Since the two layers of counter-ions are similarly charged, they repel each other. The closer the two layers approach each other, the greater is the force of repulsion. A good dispersant creates a layer of counter-ions of such thickness and concentration of electrostatic charge that particles coming together with high energy will not force the fields of repulsion of the clouds of counter-ions to merge completely; the particles will not make physical contact.

It must be remembered that the particles still have an attraction for one another when they are very close together, even in the presence of the double layer described above. This attraction, like surface tension, is the result of forces of attraction between molecules, called Van der Waals-London forces. The properties of these forces are treated quantitatively by Verwey and Overbeek *(1)*. Dispersion is made possible by the fact that the forces of attraction between particles do not become significant until the particles are very close together, that is at a separation of the order of 10 to 30 Å, whereas the forces of repulsion created by the diffuse layer of counter-ions are operative for relatively long distances (of the order of 100 Å). In the absence of a dispersant, there is no repelling force to prevent the close approach of the particles, and, when they approach closely, the forces of attraction are operative in causing flocculation, opposed only (in viscous suspension like coating colors) by the kinetic energy resulting from the motions of stirring. Even in the absence of a dispersant, particles colliding at high velocity will rebound, but high viscosity prevents the attainment of such high velocities except under conditions of agitation with high energy input.

The action of the counter-ions in creating a diffuse electrostatic field around the particles has been described. Emphasis was placed upon the adsorption of ions of the dispersant on the particle surface as the primary cause for the clustering of the counter-ions around the particle. The nature of the ions that constitute the counter-ion cloud is also important. Usually, the adsorbed ion is the anion, resulting in a negative charge on the surface of the pigment particles. The corresponding cloud of cations composing the counter-ion cloud will be positively charged. If the cations that compose the counter-ion cloud are monovalent or if the concentration of cations in the intermicellar solution is low, the cloud will be thick and voluminous, extending relatively far out into the solution from the surface of the particle. The most common monovalent cations are sodium and potassium. On the other hand, if divalent cations compose the counter-ion cloud or if the concentration of cations in the intermicellar solution is high, the cloud will be thin, hugging the surface of the pigment particle closely. Calcium ion is the most frequently encountered of the divalent ions, but zinc and magnesium may sometimes be present in significant concentrations. A thin cloud allows the particles to approach one another closely, greatly increasing the probability that the positive fields will merge far enough for attractive forces to take over, causing flocculation. This is one important reason for the destructive effect of calcium, magnesium, and zinc ions upon the stability of many pigment dispersions.

Dissolved electrolytes in the solution, such as dissolved salts in a pigment dispersion, always have the effect of collapsing the counter-ion cloud, that is, of moving it closer to the particle surface. Regardless of the dispersant used, dissolved salts, acids, or bases in high concentration will tend to reduce the stability of a dispersion.

Molecular Weight Effects of Anionic Polymers

In as much as the molecular weight of a polymer is one of its key physical properties, let us consider its importance with respect to end-use performance. Anionic polymers of low molecular weight (below 20,000) perform as dispersants. Similar anionic polymers of high molecular weight (over one million) function as flocculants. **Figure 6.4** shows that: (a) The proportion of an absorbed polymer actually in contact with the solid surface is greater if its molecular weight is low; (b) because the size of low molecular weight polymers is small in comparison with the average particle separation, multiple anchorage of small molecules occurs exclusively on the surface of a single particle; and (c) if molecular weight is increased by a factor of 10, anchorage of a single molecule on more than one particle becomes possible, forming an inter-particle "bridge."

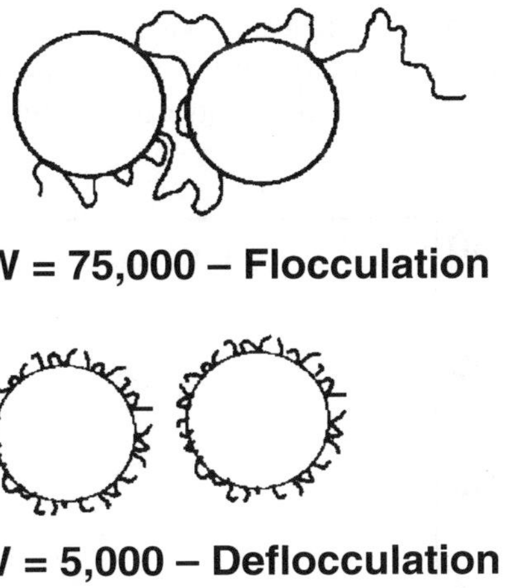

Fig. 6.4. Effect of dispersant molecular weight

As we just saw, it is extremely important to match the molecular dimensions of the polymeric dispersant to the particle size of the pigment to be dispersed. **Figure 6.5** shows how the molecular weight for optimum dispersing power varies with pigment particle size. One can see that a lower molecular weight dispersant is needed for a fine particle size pigment. Titanium dioxide (100% less than 2 microns) requires a lower molecular weight dispersant than kaolin (80% less than 2 microns). A much higher molecular weight dispersant is required for dispersing calcium carbonate (31% less than 2 microns).

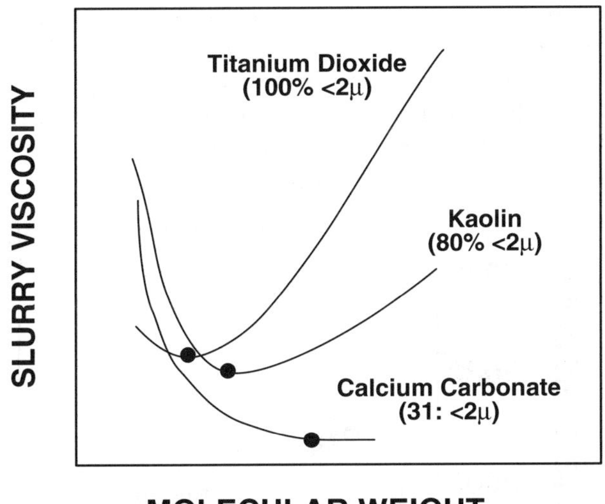

Fig. 6.5. Dispersing power vs. molecular weight

Electrolyte Effects on Dispersant Performance

All electrolytes compress electrical double layers and eventually reduce the surface potential itself, both effects reducing any electrical barriers to flocculation. Furthermore, they will cause absorbed polyelectrolytes to adopt a more tightly coiled configuration, thus adversely affecting the steric barrier. Salts of multivalent cations can, in addition, exert a special effect on polycarboxylates through their ability to "bind" carboxyl groups.

Although the level of Ca++ present in hard water usually has negligible effect on dispersant performance, higher levels can cause crosslinking and insolubilization of the polymer. The effect of simple electrolytes on dispersing power of dispersants is illustrated in **Fig. 6.6**.

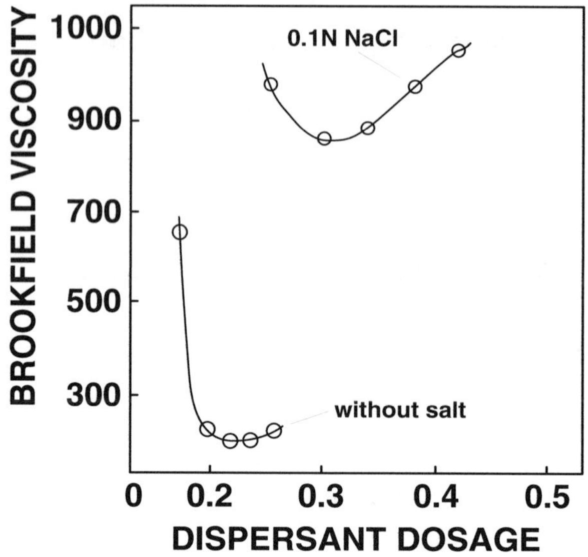

Fig. 6.6. Added electrolyte effect on slurry viscosities

The effect of polyvalent cations, specifically Al+++ in this example, is illustrated in **Fig. 6.7**. Tests revealed that filtrate from the unwashed slurry had high electrical conductivity and that high concentrations of Al+++ were present. The slurry was washed with an excess of tap water, filtered, and restored to the original solids content before the dispersant was re-evaluated. The performance of the sodium polyacrylate was dramatically improved, since it was previously extremely vulnerable to binding by Al+++ and to the compressing effect of all electrolytes.

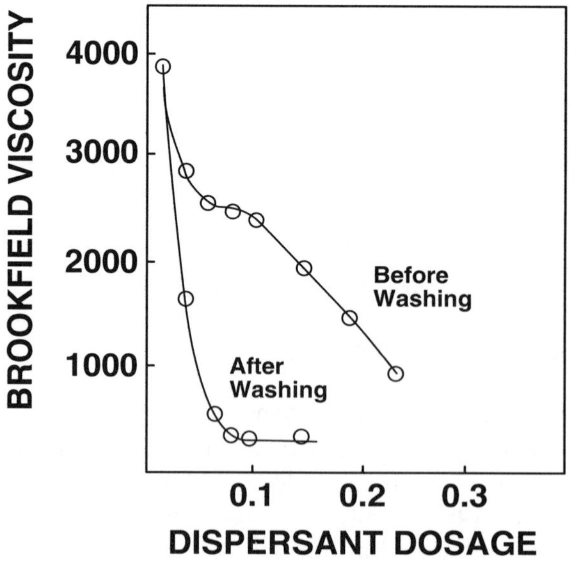

Fig. 6.7. Effect of electrolyte removal

pH Effects on Dispersants

Increasing the pH of a mineral slurry: (a) will increase the degree of ionization of carboxylic polymers; (b) will increase the negative charge of the underlying mineral surface in many cases (Both of these trends would increase the total charge carried by the particles if absorption was unchanged. However, adsorption of dispersants may either increase or fall with pH; e.g., adsorption of polyacrylic acid on $CaSO_4$ increases with pH; adsorption of polyacrylic acid on kaolinite falls with pH.); and (c) addition of base to the system to increase pH will raise the ionic strength, eventually having an adverse effect on the dispersion stability.

In general, increasing either pH or dispersant dosage will at first improve deflocculation, but large increases in either may have an adverse effect because of the accompanying increase in ionic strength, thus large dispersant doses might be expected to be most effective at lower pH, small doses at high pH. **Figure 6.8** shows that this is so in practice, a viscosity of 300 cP being achieved by only 0.10% of sodium polyacrylate at pH 7.50, but requiring a dose of 0.1% at pH 5.57. Nevertheless, the lowest viscosities of all were obtained at the lowest pH, although requiring high dosage of dispersant to achieve them. Although this appears to be a costly way of achieving a given viscosity reduction, the alternative approach of using less dispersant and adjusting the pH upward could result in less stable viscosity and, possibly, inferior flow properties under different conditions.

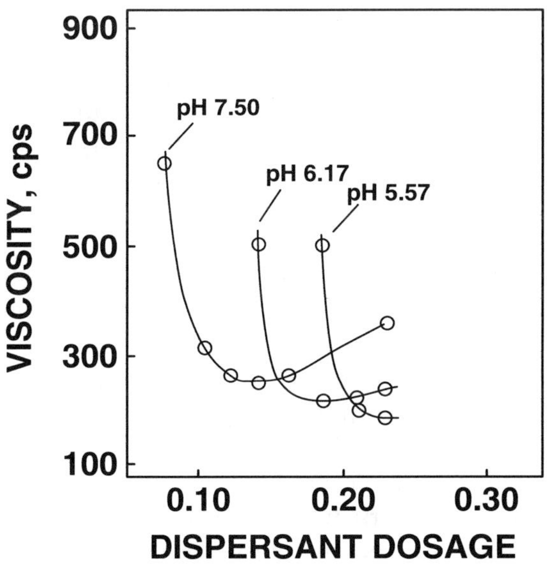

Fig. 6.8. pH effect on dispersant efficiency

The Polyphosphates

The polyphosphates are the second most important class of dispersants used in preparing aqueous pigment dispersions or paper coating colors. The fundamental building blocks of the polyphosphates are phosphorus-oxygen tetrahedra polymerized into chains of various length. The practical building blocks of the polyphosphates are the monosodium and disodium salts of orthophosphoric acid. The polyphosphates are formed by the molecular dehydration of these salts or mixtures of them.

Tetrasodium pyrophosphate, or sodium diphosphate, is formed by dehydrating two moles of disodium orthophosphate:

$$2Na\,HPO_2 \longrightarrow Na_4P_2O_7 + H_2O \tag{1}$$

Pentasodium tripolyphosphate, or sodium triphosphate, is formed by dehydrating two moles of disodium orthophosphate and one mole of monosodium orthophosphate:

$$2Na_2HPO_4 + NaH_2PO_4 \longrightarrow Na_5P_3O_{10} + 2H_2O \tag{2}$$

Sodium tetraphosphate is formed by dehydrating two moles of monosodium orthophosphate and two moles of disodium orthophosphate:

$$2NaH_2PO_4 + 2Na_2HPO_4 \longrightarrow Na_6P_4O_{13} + 3H_2O \tag{3}$$

Sodium metaphosphate is formed by dehydrating monosodium orthophosphate:

$$xNaH_2PO_4 \longrightarrow (NaPO_4) + xH_2O \tag{4}$$

The degree of polymerization, or chain length of the metaphosphate will depend, in part, upon the dehydration temperature, the conditions under which the molten material is cooled, and the mole ratio of Na_2O to P_2O_5. Many commercial metaphosphates are produced with a mole ratio of Na_2O/P_2O_5 slightly greater than 1.0. This has the effect of lowering the degree of polymerization and increasing the dissolving rate because of lower solution viscosity. The pyrophosphates and tripolyphosphates are crystalline. The tetraphosphates and metaphosphates are amorphous glasses; the metaphosphates, however, can exist in several crystalline forms that have no utility in paper coating. The glassy phosphates usually are manufactured with P_2O_5 contents of 61.5 to 67%; their chain lengths vary from 4 to 20 phosphorus atoms per chain.

The crystalline phosphates are usually made by adding the correct proportion of soda ash and phosphoric acid to a mixing tank, drum-drying the resulting slurry at a temperature below that required for molecular dehydration, and calcining in a suitable kiln to allow the resulting orthophosphate to achieve molecular dehydration. The product is then cooled, crushed, and packaged.

In the manufacture of the glassy phosphates, the orthophosphate slurry is fed directly into a shallow gas-fired furnace where molecular dehydration takes place and molten glass is formed. The melt is allowed to flow out of the furnace onto a chilling wheel where the glass shatters into small fragments. The fragments are either packaged directly or crushed to a granular form before packaging, depending upon the particle size desired in the finished product.

The amount of polyphosphate required for optimum dispersion varies with the surface area as well as the chemical structure of the particular pigment being used. Typical levels of addition vary from a few tenths of one percent to one percent or more based on the weight of pigment.

The structure of the polyphosphate anions is such that they are fairly efficient sequestering agents for polyvalent cations. Since polyvalent cations in solution are undesirable in effecting maximum pigment dispersion, and since these ions are frequently present in the makeup water available to paper mills, the sequestering action of the phosphates is frequently a significant reason for their use as dispersants in practical mill situations.

While many of the polyphosphates are excellent dispersants, many of the simpler types of phosphate compounds are relatively ineffective. The orthophosphates from which the polyphosphates are derived, for example, fall into this category. Unfortunately, just as the polyphosphates are formed from orthophosphates under appropriate reaction conditions, so the polyphosphates under other conditions will hydrolyze back to the orthophosphates. Green *(2)* had shown the elevated temperatures and/or acid conditions increase the rate of hydrolysis (reversion) of the polyphosphates. Because of low cost and effectiveness, tetrasodium pyrophosphate is by far the most widely used phosphate dispersant. However, Green had shown that this phosphate is more subject to hydrolysis than the other polymeric or condensed phosphates he evaluated.

This reversion phenomenon represents one of the greatest commercial limitations to the use of polyphosphate dispersants, especially under conditions of long-term slurry storage, or where predispersed spray-dried pigments may be stored in silos or railroad cars at elevated temperatures for long periods of time. Reversion of the pyrophosphate to the orthophosphate form results in the release of hydrogen ions so that the rate of reversion increases once hydrolysis commences. In some cases, the hydrolysis effect on rheology can be minimized by the addition of increments of fresh polyphosphate dispersant. Efforts along this line do not result in a return to the viscosity of the original, freshly prepared slurry because of the increased salt content that is then introduced into the suspension.

The most common solution to the phosphate-reversion problem is to select conditions in which the reversion reaction is less favorable and chemically stable, or that derives its dispersing ability from a protective-colloid mechanism that will subsequently be described.

The polyphosphates are especially effective for dispersing clays *(3)*, calcium carbonates *(4)*, and titanium dioxide *(5)*. The dispersion of calcium carbonate presents a unique problem in that this pigment becomes soluble under acid conditions, and even at neutral pH some divalent soluble calcium ions can be expected. Since these have a profound undesirable effect upon suspension stability, those polyphosphates that possess the greatest sequestering power tend to be the best dispersants for calcium carbonate slurries *(4)*.

The Alkali Silicates

The fundamental building blocks of this class of dispersants are silicon-oxygen tetrahedra linked together in chains and networks. In this respect, they exhibit a structure analogous to the polyphosphates. The alkali silicates are formed by fusing soda ash and sand in various proportions as follows:

$$Na_2CO_3 + nSiO_2 \longrightarrow Na_2O \bullet nSiO_2 + CO_2 \qquad (5)$$

The value of n for water-soluble sodium silicates can vary from values less than unity to about 3.5. Silicates containing certain specific ratios of Na_2O/SiO_2 have definite names. Thus, for N = 0.5, the compound is known as sodium ortho silicate; for n = 0.67, sodium sesquisilicate; for n = 1.0, sodium metasilicate; and for n = 2.0, sodium disilicate. Aqueous solutions of these silicates have been known generally as "water glass."

In the manufacture of sodium silicates, soda ash and sand are melted together in an open-hearth type of glass-melting furnace at a temperature of about 1450°C. The melt can be dissolved directly from the furnace in rotary dissolvers or chilled, crushed, and dissolved under pressure with steam. Sodium sesquisilicate and sodium metasilicate can exist in crystalline forms and are frequently sold as dry solids.

As dispersants for most pigment systems, the alkali silicates are less efficient than the polyphosphates and tend also to be most effective at somewhat higher pH values. This is probably because the silicic acids are ionized to a lesser degree at neutral pH than are the polyphosphates. In spite of this limitation, in some applications, the alkali silicates are preferred as pigment dispersants because they tend to be less strongly adsorbed on the mineral surface and hence do not cover up the unique desired functional properties of the pigment surface. Just as was the case with the polyphosphates, the alkali silicates are effective dispersants for clays, calcium/carbonates, and TiO, although they are less generally used than the phosphates in paper coating systems.

The Alkalies

The alkali-metal hydroxides or carbonates are frequently used as part of the pigment-dispersant system, but are rarely used as the sole dispersant. In most cases, they function merely as pH adjusters since the commonly used dispersants function most effectively under neutral or slightly alkaline conditions. When used by themselves, these dispersants have several serious disadvantages:

1. They function most effectively at pH values sufficiently high (pH 10 and above) that some pigments become chemically reactive. Because of the high permanent (nonvolatile) alkalinity required for maximum dispersion, dried coatings dispersed entirely with alkalies may cause problems of ink drying following the printing operation.
2. The levels of addition required to provide maximum dispersion (that is, neither underdispersed nor overdispersed) are so critical that problems of control arise when attempts are made to use them commercially as the sole dispersant.
3. Since they possess no sequestering power, maximum pigment dispersion cannot be obtained unless the water used in the makedown is substantially free of polyvalent cations. In addition, they are ineffective in calcium carbonate systems because of soluble calcium ion interference, although they can be used to disperse clay or titanium dioxide.

   In general, the carbonates are less effective as dispersants than are the hydroxides, although they are frequently used as dispersants for "in process" dispersion of pigments where free-flowing slurries at relatively low solids content are desired. More recently, the alkali-metal fluorides have also been found to function as effective pigment dispersants for slips prepared under acid conditions, although these are of little or no commercial significance.

The Nonionic Polymer Dispersants

In addition to the conventional dispersants already discussed, there are a large number of low-molecular-weight, nonionic polymers and/or surfactants that also exhibit a deflocculating action. Most common of these is a family of condensation products of fatty alcohols and ethylene oxide *(6)*. In this case, the more hydrophobic end of the surfactant molecule attaches to the pigment surface, while the hydrophilic polyoxyethylene chains project into the aqueous phase. These projecting chains produce a layer of sufficient mechanical strength around the particles so that the collisions that occur between particles do not permit the closeness of approach necessary for flocculation to occur. This has been termed a "protective-colloid" phenomenon. Details by which this protective-colloid action comes about are not yet clearly understood, although an interesting theoretical treatment of the problem has been provided by Clayfield and Lumb *(7)*. Results of their calculations suggest that there is an optimum ratio between the chain length of the polymer and the size of particle that can be stabilized. Their work also indicates that, the greater the freedom of movement of a polymer as long as it remains adsorbed on the particle surface, the more effective it will be as a dispersant.

Crowl *(8)* has developed an expression in which the energy of interaction between two particles can be predicted by taking into consideration the presence of an adsorbed layer of polymer on the particle surface. His conclusions indicate that the stabilization effect of a 50 Å thick adsorbed layer is greater on a suspension of 1.0 μ particles than on a suspension of 0.2 μ particles. He also concludes that stability of a pigment system by stearic hindrance (protective-colloid action) is greatest if the adsorbed polymer chains form a close-packed, incompressible layer with a high energy of desorption of the adsorbed layer.

While the protective-colloid mechanism of dispersant action is probably operative to some degree in all dispersed pigment systems, there are some dipsersants in which this mechanism appears to be the dominant effect. Thus, starch and carboxymethylcellulose have been used to disperse calcium carbonate pigments; gum arabic has been used with some effectiveness as a dispersant for TiO; and derivatized starches, proteins, and gums have been used to disperse satin white *(9)*.

An excellent review by Athey *(10)* provides a description of the many types of organic dispersants that have been used or suggested for the dispersion of paper coating pigments. Also included are a discussion of theoretical aspects of pigment dispersion and of methods for evaluating dispersants. Morris and Hochenedel *(11)* provide a good review of pigment dispersants as well as a comparison of polyphosphate and polyacrylate dispersants with respect to particle charge, rheology, and thermal stability effects. Other interesting comparisons of the polyphosphate with organic polymer-type dispersants have been give by Greene and Reder *(12)*, Hern and Fritz *(13)*, and Rohmann *(14)*.

At the beginning of this chapter, it was assumed that the pigment particles were wetted by the suspending medium. Hence, our concern herein has been with dispersion and stabilization rather than with wetting phenomena. The mineral talc is the single white pigment of some commercial significance in paper coating that is not spontaneously wetted by water. In dispersing this mineral, it is therefore necessary to add a surfactant to first provide complete wetting, and then to add a polyphosphate to provide the dispersant action *(15)* necessary to permit the preparation of relatively high-solids talc suspensions.

Formulating with Dispersants

Figure 6.9 is an example of a pigment dispersant demand curve. As dispersant dosage is increased: (a) viscosity decreases rapidly at first, (b) a viscosity minimum occurs at optimum dosage, (c) viscosity gradually increases beyond optimum dosage. The rapid viscosity drop (a) is caused by the increase in surface potential and steric barrier as the dispersant is adsorbed on the particles. The viscosity minimum (b) signals the optimum dosage of dispersant for the specific system and conditions. The gradual rise beyond optimum dispersant dosage (c) is related to an accompanying adverse effect of increasing counter-ion concentration (NH+ in this example).

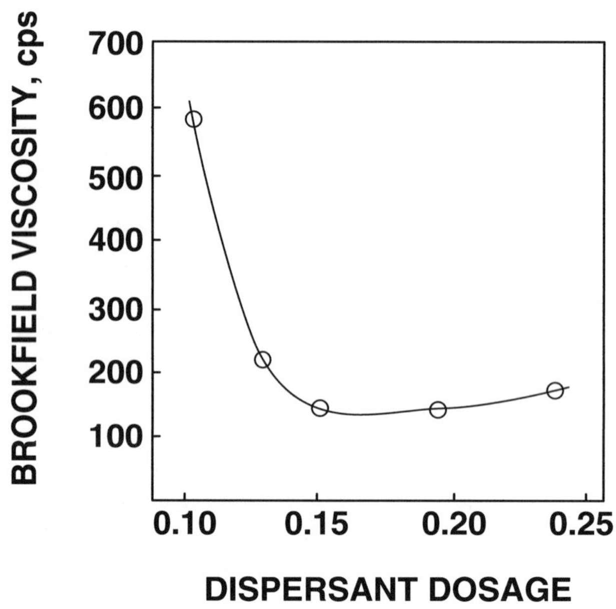

Fig. 6.9. Typical dispersant demand curve

The dispersant demand curve evaluation is a preliminary evaluation tool for measuring dispersion of a system: Additional evaluations for storage and aging stability, and other rheological properties are recommended. **Figure 6.10** is an example of Hercules High Shear Rheograms of a No. 1 grade kaolin slurry.

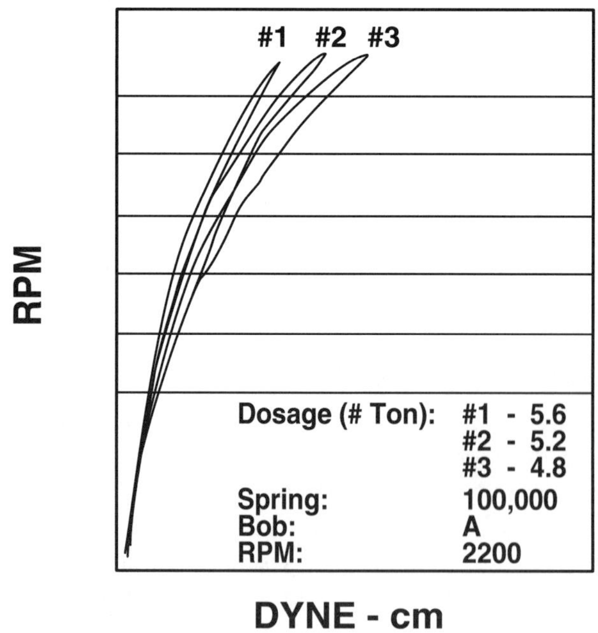

Fig. 6.10. Hercules HSV data on dispersant demand

Literature Cited

1. Verwey, E.J.W. and Overbeek, J. T., in *Theory of the Stability of Lyophobic Colloids*, Elsevier Publishing, New York, 1948.
2. Green, J., *Ind. and Eng. Chem.* 42: 1542–1546(1950).
3. Asdell, B. K., *Paper Mill News* 70: 82–96(May 31, 1947).
4. Leaf, C.W. and Liggett, L.M., *Tappi* 39: 142-7(1956).
5. Bowman, A. and Hughes, W. J. *Oil Colour Chem. Assoc.* 34: 412–442(1951).
6. Carr, W., *J. Oil Colour Chem. Assoc.* 54: 155–173(1971).
7. Clayfield, E. J. and Lumb, E. C., *J. of Coll. and Interface Sci.* 22: 269–284(1966).
8. Crowl, W. T. and Malati, M. A., *Discuss, Faraday Sec.* 42: 301–312(1966).
9. Thompson, R. N., U.S. Pat. 2, 671,032 (1954).
10. Athey, R. D. Jr., "Polymeric organic dispersants for pigments: colloid science and practice (part 1)," *Tappi* 58(9): 66–71(1975); "Polymeric organic dispersants for pigments: useful structures and their evaluations" (part 2), *Tappi* 58(10): 55–61(1975).
11. Morris, H. H. and Hochenedel, J. L., "Pigment and Coating color Dispersants," presented at TAPPI Pigments Workshop, Philadelphia, 1975.
12. Greene, G. W. and Reder, A. S. "Electrokinetic and rheological properties of calcium carbonate dispersions used in paper coating," *Tappi* 57(5): 101–106(1974).
13. Hern, J. F. and Fritz, J. H., "Auxiliary dispersants for pigments and pigmented coatings," *Tappi* 49(12): 77–88A(1966).
14. Rohmann, M. E., "The influence of sodium polyacrylate on the properties of coating colors," *Tappi* 57(5): 96–100(1974).
15. Lamar, R. S., "Talcs" in *Paper Coating Pigments,* TAPPI Monograph No. 20, TAPPI, Atlanta, 1958, Chap. 10, pp. 157.

Bibliography

Athey, R. D. Jr., "What are the pitfalls in pigment dispersion formulation?," *Paper Trade J.* 159(12): 56–57(1975).

Buechler, P. R., Borwn, G. L., Parikh, V. P., and Salmon, H. J., *J. of Paint Tech.* 45: 60–64(1973).

MacGugan, I. C., *Tappi* 49(7): 61A–62A(1966).

Pascal, R. H. and Reig, F. L., *Official Digest* 36: 839–52(1964).

Van Essen, W. J., Erickson, D. E., and Rolik, M. A., "High-efficiency pigment binders based on hydrophilic modifications of styrene-butadiene copolymers," *Tappi* 50(12): 622–29(1967).

Zisman, W. A., Chap. 1 in *Contact Angle, Wettability and Adhesion,* No. 43 in Advances in Chemistry Series, ACS, Washington, DC, 1964.

Acknowledgment

The author is indebted to G. A. Hemstock, author of "Dispersants," Chapter 2 in the previous coating additives monograph (*Paper coating additives,* TAPPI PRESS, Atlanta, 1976, pp. 15–33). He has borrowed liberally from this work.

CHAPTER 7: LUBRICANTS

John M. Krasniewski, Jr.

Introduction

Coating lubricants were initially used in the early paper coating colors that used bulky kaolin clay pigments imported from Great Britain with binders such as animal glue or casein. Typically, these coating colors were applied at relatively low solids with a slow-speed, roll-coating application. The majority of the paper being manufactured was designed to be used in letter press printing. After drying, the coating was relatively brittle due to the use of the natural binders. In order to reduce coating cracking and pigment dusting during the finishing, slitting, sheeting, and printing operations, it was necessary to identify materials that would plasticize the coating. Waxes and metal soaps were found to fulfill this need. When these additive materials were used, finished sheet print smoothness was also improved due to better pigment orientation through easier binder flow. Those who made coated paper detected a "change" in the wet coating color that contained a lubricant. The colors were easier to apply and meter. It was necessary, then, to *lubricate* the coating in order to impart a desired wet state flow property.

Modern, high-speed printing processes have placed numerous demands on the performance of coated paper. These demands have necessitated changes in the pigments and binders as well as lubricants. Coating colors are now applied via long- or short-dwell time type applicators. The desired coating weight is achieved via blade-metering techniques. While the printer still requires efficiency, economy, and high quality from the manufacturers of coated paper, letterpress printing has given way to offset lithographic as well as rotogravure printing techniques. Along with the required optical properties of brightness and opacity, minimum surface pick strength and water resistance characteristics are major requirements for coated papers that are to be printed via offset lithography. For rotogravure printed sheets, print smoothness is a major property concern. The use of synthetic latex binders provides for greater binder flow in order to meet the ever-increasing print smoothness and sheet gloss requirements.

In the offset grades, the coating color binders for the No. 1 through No. 3 heavyweights typically contain only the synthetic, latex products. Those used for No. 4, middle weights, are usually low starch/high latex blends. For No. 5, LWC coating colors, the binder may consist of 50% starch and 50% latex. Some colors for No. 5 grades may approach 75% starch and 25% latex. Pigment binders containing soy-protein/latex combinations are found in the majority of paperboard coatings as well as some publication grades. Paper machines now exhibit paper trim widths of >7.5 meters as coating speeds of >1000 meters per minute are typical.

The pigments used today include a variety of mechanically and chemically treated kaolin clays. With the renewed interest in the *alkaline* papermaking process, the use of calcium carbonate pigments, precipitated on the mill site or mechanically ground from natural sources, is growing in popularity. All of these ingredients are designed to provide optimum coating performance to both the wet state and dry coated surface.

The maker of coated paper now is concerned with coating immobilization *(1)* or the location on the surface of the rawstock where flow ceases after metering as well as the manner in which the coating interacts with the rawstock. The origin of the fiber and the manner in which it is made into pulp has been identified a major consideration in the nature of the surface characteristic of the finished sheet *(2, 3)*. The use of pigment reactive latices *(4)* and modified kaolin clays *(5)* provide the papermaker with a means of achieving an enhanced, structured coating matrix which provides print smoothness that could not be achieved in the days of roll coaters and casein binders.

With all the advances in coating color materials and the new approaches in identifying and defining the fundamentals of the manufacture of coated paper, coating lubricants are still used and they continue to perform a major role in the manufacture of coated grades. Lubricants are still needed to prevent pigment dusting during the finishing, slitting, sheeting, and printing operations. Pigment orientation as a result of binder flow is easier to achieve when a coating lubricant is present. A desired print smoothness is easier to achieve with a coating lubricant. Coating color viscoelastic character remains a concern because coating lubricants still influence the wet state character of the coating color so that it is "changed" for better application and metering. The surface character of the dry state coating is altered when a coating lubricant is present such that the finished sheet properties are optimized.

While the need for a coating lubricant is generally accepted, identifying its exact function can be a difficult — if not impossible — task. In most instances, the lubricant function will differ according to the requirements of the grade and the habits of the paper mill. It is difficult to judge the performance, let alone the need for a lubricant, as a separate entity. Its benefit must be appraised in terms of some property of the finished sheet or the manufacturing process.

This treatise will discuss the chemical products used today as paper coating lubricants. An attempt will be made at establishing the functional contribution in the complex coating color systems resulting from the use of these lubricant materials. The recognized complexity of a coating color system will necessitate that a phenomenological approach be used such that conclusions are limited to strong indications or beliefs based upon empirical data and intuition. Sufficient knowledge exists to at least understand the behavior of coating lubricants in coating colors even though the benefits cannot easily be explained.

Types of Coating Lubricants

Hern *(6)* has categorized by chemical composition the materials that may be useful as paper coating lubricants. While all of the materials listed by Hern may still provide a lubricant's function in today's high-speed manufacture of coated paper and paperboard, the products presently used fit into the following groups:

1. Aqueous dispersions of metallic soaps
2. Wax emulsions
3. High-solids, lubricant blends containing calcium stearate
4. Soy-lecithin/oleic acid blends.

While specialty mineral oil/sulfated petroleum oil blends as well as a variety of natural and synthetic wax emulsions are available in the marketplace, these products are designed to be used in nonpigment-containing systems for uncoated papers and paperboard that require unique surface slip characteristics.

Metallic Soaps

Metallic soaps are formed by the reaction of a metal alkali or oxide with a fatty acid. The most commonly used metallic soap is the calcium salt of stearic acid, calcium stearate. In the manufacture of calcium stearate, the typical calcium source is CaO (*quicklime*) that has been hydrated in order to make the reactive base, $Ca(OH)_2$. Commercially available stearic acid ($C_{18}H_{36}O_2$) is produced from animal, marine, and/or vegetable fats and oils. It is not pure stearic acid. Depending upon the source of the fat and the method of production, *stearic acid* may contain large amounts of other organic acids ranging from lauric acid ($C_{12}H_{24}O_2$) to behenic acid ($C_{22}H_{44}O_2$) or unsaturated acids such as oleic or linoleic. Accordingly, a stearate may contain significant amounts of laurate, palmitate, or oleate.

For the calcium stearate that is supplied as a paper coating lubricant, the stearic acid source is normally beef tallow. The usual fatty acid molecular weight distribution is: 65% stearic ($C_{18}H_{36}O_2$), 29% palmitic ($C_{16}H_{32}O_2$), 4% myristic ($C_{14}H_{28}O_2$), 1.5% margaric ($C_{17}H_{34}O_2$), and 0.5% pentadecanoic ($C_{15}H_{30}O_2$) along with trace amounts of the unsaturated fatty acids, oleic ($C_{18}H_{34}O_2$) and palmitoleic ($C_{16}H_{30}O_2$) *(7)*. A calcium soap from this fatty acid mixture will exhibit a melting range of 245–285°F. **Figure 7.1** is a DSC curve for a calcium soap made from tallow fatty acid.

SAMPLE: *Calcium Stearate Powder*	SIZE: *0.78 mgm.*	ATM: *Nitrogen @ 30ml./min.*	RUN NO: *1716*
	REF.: *Empty Pan*		DATE: *1/5/81*
	PROGRAM MODE: *Heating*	SCALE SETTING: *20°/in* *y = 0.1 mcal/sec./in.*	OPERATOR: *R. C. Klepper, Jr.*
ORIGIN: *J. Krasniewski*	RATE: *5°/min.* START: *Room Temperature, °C*		

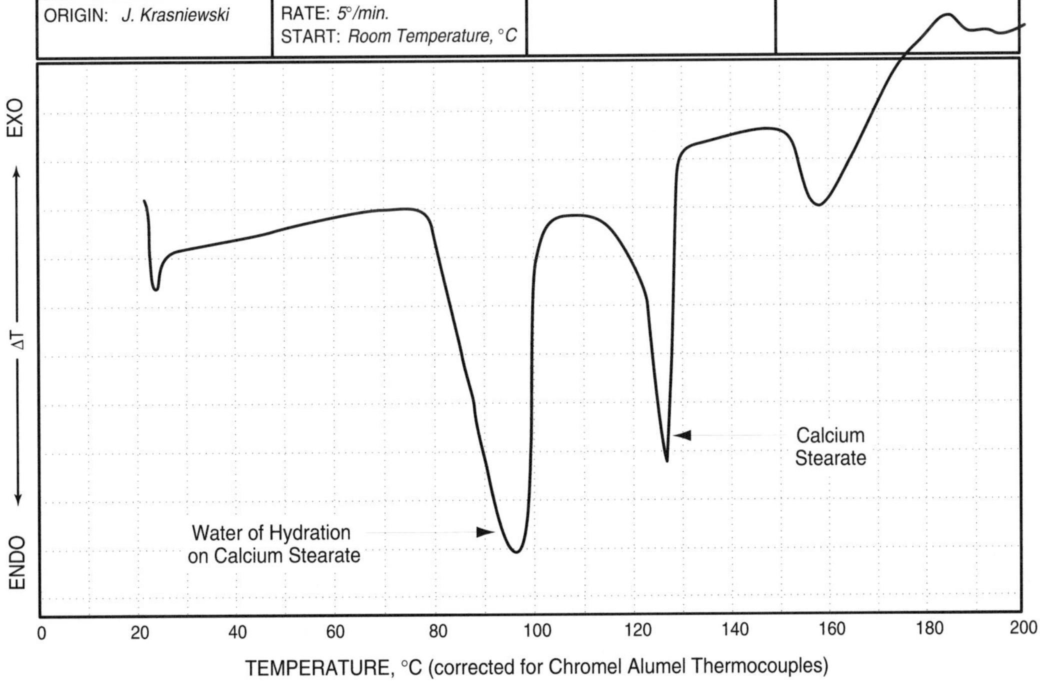

Fig. 7.1. DSC curve for a calcium soap made from tallow fatty acid

Manufacture of Calcium Stearate Dispersions

Today's typical coating speeds of >3000 ft/min demand that significant volumes of coating color be delivered to the coating station in a timely and efficient manner. This has led to the development and popularity of the *continuous batch*-type coating makedown system which makes this possible. In this system, liquid ingredients can be metered into the process with positive displacement pumps through the use of magnetic flow meters. In order to meet this requirement, calcium soaps used as paper coating lubricants are supplied as aqueous dispersions with total solids of 50–55%. As dry calcium stearate is insoluble in water, a dispersion is best made by manufacturing the soap in water that contains an appropriate dispersant *(8)*. Normally, the calcium soap is made in an aqueous system containing 3–5% nonionic and/or anionic dispersants. In order to ensure extremely low retain residue on a U.S. Standard 325 mesh screen as well as a uniform particle size distribution, the calcium soap dispersion is subjected to a homogenization and/or milling process. Screen retention of <50 ppm are typically attained. While milling is also important for attaining a desired particle size distribution, the nature of the distribution is also dependent upon the manner in which the calcium hydroxide is combined with the tallow fatty acid. If the calcium hydroxide is added to an aqueous dispersion of the tallow fatty acid, a rather coarse particle size distribution is achieved. Those who manufacture the product usually call this a *standard* addition process. If the *reverse* addition process is used, whereby the tallow fatty acid is added to an aqueous slurry of calcium hydroxide, a relatively fine particle size dispersion is achieved. The significance of these physical property differences in particle size distribution will be discussed in the section concerned with the function of calcium stearate and coating color viscoelastic character. **Figure 7.2** differentiates the particle size distributions for calcium stearate dispersions made by the *standard* and *reverse* addition processes. These distributions were generated with a Coulter LS 130 laser light-scattering particle size analyzer.

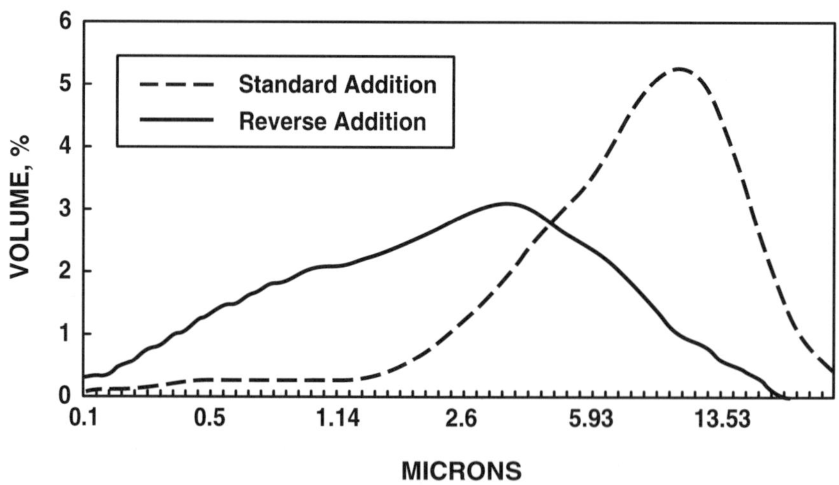

Fig. 7.2. Particle size distributions for calcium stearate dispersions

Calcium Stearate Dispersion Handling and Addition

Calcium stearate dispersions are extremely shear stable and exhibit relatively low viscosity. They are designed to meet the metering requirements for continuous makedown systems. Magnetic metering can be easily employed. The typical conductivity for a calcium stearate dispersion is approximately 20 micro mhos/cm.

The amount of calcium stearate used in the coating color can vary from mill to mill and grade to grade. A normal usage level is 1.0% dry lubricant on dry pigment. Amounts as much as 1.25% have been used in grades requiring critical surface character conditions. The product should be thoroughly mixed directly with the coating color at any convenient point in the coating makedown.

Calcium stearate dispersions are pH sensitive. They are stable in coatings at pH levels from 6.0 to 10.5. High acid pH coatings should be adjusted to this pH range prior to addition. Calcium stearate dispersions are reasonably stable in the presence of alkalis. This product should not be subjected to acid solutions, since decomposition of the calcium stearate will result.

Freezing will seriously damage aqueous dispersions of calcium stearate. Bulk storage tanks in severe climates should be fitted with a submerged tank heater operating on controlled low pressure steam.

Wax Emulsions

Aqueous emulsions of waxes have been available as coating lubricants almost as long as coated paper has been manufactured. For surface sizing with starch or other film forming materials, the inclusion of a wax emulsion in the size solution will improve sizing, smoothness, scuff resistance, printability, surface slip and fold. The wax materials used to make the aqueous emulsion exhibit melting points that range from liquid hydrocarbon oils to high melting point (>140°C) oxidized polyethylene. Wax emulsions are available in several tailored forms and may be prepared from a wide range of waxes selected to fulfill the numerous requirements of the coated paper and coated paperboard manufacturers. The groups of waxes that form the basis of these products are paraffin waxes, microcrystalline waxes and slightly oxidized, low density to high density polyethylene waxes. Special types, grades and blends of waxes are often specially manufactured as emulsions in order to meet a mill's particular requirements.

1. Paraffin Waxes

Commercial paraffin waxes consist of straight chain or n-paraffin hydrocarbons of the C_nH_{2n+2} series. Average molecular weights are in the range of 360 to 420 with melting points of 115°F to 150°F. Higher melting point (160°F) paraffins have molecular weights up to 600. Fully refined paraffin waxes are fairly hard, white in color and contain a maximum of 0.5% oil.

Semi-refined paraffin waxes vary in color, hardness and oil content. Compared to the refined paraffins, the color may vary from light amber to off-white, the penetration value is slightly higher and the oil content is within the range of 1% to 5%. For many grades of paper and paperboard, the semi-refined paraffin waxes impart the required properties.

2. Microcrystalline Waxes

Microcrystalline waxes are primarily alkylated naphthenes containing side chains, isoparaffins, some n-paraffins and a small amount of alkylated aromatics. Average molecular weights are within the range of 580 to 700. A few special micro waxes have molecular weights as high as 900. Melting points range from 130°F to 200°F. Micro waxes have considerably smaller crystals than paraffin waxes. They offer considerably different physical properties. Micro waxes are more flexible, have greater tensile strength and generally have a greater degree of tackiness when compared with paraffin waxes. This latter property decreases the slip of micro wax treated paper and paperboard when compared to a paraffin wax treated surface.

3. Oxidized Polyethylene Waxes

Oxidized polyethylene waxes are made from petroleum-based polyethylene that has been slightly oxidized in order to achieve better emulsion stability. They are readily emulsified with an added surfactant, followed by the addition of caustic to a pH of 9.0. They offer a wide range of property improvements including improved slip, scuff resistance along with the typical barrier properties associated with the use of a wax. In pigment coating applications, the use of an oxidized polyethylene reportedly gives good coating plasticizing and fold properties *(9)* as well as good glossing, but it is less effective than metallic soaps as an antidusting agent. This may be due to the rather sharp melting point exhibited by the oxidized polyethylene wax. The solids content for the aqueous emulsions products range from 25% to 60%. The higher solids products contains fatty acid esters of ethylene oxide and urea in addition to the polyethylene wax. While the materials used in the aqueous emulsions are considered to be low molecular weight waxes, their physical properties are defined by the nature of their "density" or gm/cc. A typical low density, oxidized polyethylene exhibits a density of 0.93 gms/cc and a melting point of 104°C. A high density analogue exhibits a density of 0.98 gms/cc and a melting point of 140°C.

In applications for uncoated paper and paperboard, dilutions of oxidized polyethylene emulsions have been applied at the size press or through surface spraying in order to lower the surface coefficient of friction. Such a unique application in the manufacture of corrugating medium, where the spray application of the oxidized polyethylene emulsion has dramatically decreased medium cracking during the corrugation step.

Lubricant Blends Containing Calcium Stearate

Recently, high-solids coating lubricants based upon calcium stearate dispersions have been developed *(10–12)*. Typically, the products contain differing amounts of calcium stearate, a fatty acid ester of a polyethylene glycol and urea. Sulfur-containing compounds *(11)* as well as polysaccharides *(12)* may also be present. All of these products exhibit a total solids content of 65%.

These lubricants were designed to replace the traditional calcium stearate dispersions as they provide freight savings by offering higher total solids than the typical calcium soap dispersion. Their manufacturers claim that they also provide lubricant functionality that is typical to the calcium stearate product. The performance of these lubricant blends should be evaluated according to the standards set for each coating application and mill.

Soy–Lecithin/Oleic Acid Blends

The newest materials being supplied as coating lubricants are the soy-lecithin/oleic acid blends *(13)*. The use of these diglyceride products reportedly provides for outstanding coating rheological characteristics such that coating blade weeping and whiskering are eliminated. Higher coating solids can be used with minimum blade scratching and improved gloss ink holdout. The products are available as 50% and 100% active solids materials. The 100% active product exhibits a very low conductivity. Therefore, its use requires a positive displacement meter or mass flow meter for the introduction of measured amounts during coating makedown. While the high-solids product contains no water and will not be damaged by freezing temperatures, it will thicken significantly at temperatures below 55°F. The 50% solids product also contains urea along with the soy-lecithin and oleic acid.

The Nature of Coating Color Lubrication

Coating colors contain pigments and an adhesive or binder which are uniformly dispersed in an aqueous liquid phase. Many demands are made on this system during application and metering so that additive materials are required. This requirement makes the system very complex. The coating lubricant, in satisfying performance demands, adds to the complexity of the system.

The latest edition of the TAPPI PRESS report *(14)*, "Paper coating additives, Description of Functional Properties and List of Available Products," defines *lubricity* as a "change in the smoothness and flow properties of the wet coated surface and enhancement of the plasticity and slipperiness of the dried coating by altering the surface and/or interfacial tension of coating colors." From this definition, lubricants provide two meaningful functions for the grade of paper being manufactured. They provide an alteration and subsequent improvement to the wet state coating color flow properties along with an influence on the surface character of the dry coating matrix. For the wet coating color, the viscoelastic character is influenced by the manner in which the various ingredients of a coating color interact with one another. For the dry coating, the achievement of desired optical properties for the finished sheet are a result of the manner in which the pigment particles and other particulate materials are aligned with each other after the drying and supercalendering steps.

Wet State Coating Lubrication and Viscoelasticity

An approach for studying a complex system to identify the role that each of the components plays is to mathematically separate the effect of each of the components and then reassemble the total relationship. For ingredients in a coating color, it will be necessary, then, to start with the continuous, liquid phase of a coating color — water — and build from it.

A liquid is defined *(15)* as a substance with a fixed volume which, when under constant temperature and pressure, will assume the shape of the container in which it is held. The liquid flows in order to conform to the shape of its container. The flow follows the following relationship:

$$\frac{F}{A} = \eta \frac{dv}{dx} \tag{1}$$

where:

F	=	the force applied
A	=	the area to which the force is applied
η	=	the viscosity coefficient
F/A	=	the force required to cause flow; stress
dv/dx	=	the shear rate, an indication of the amount of flow.

This relationship indicates that any increase in force will result in a proportional increase in flow. This is the well known Newtonian flow relationship and holds quite well for water, the continuous phase of a coating color.

The viscosity coefficient is a measure of the functional forces between the adjacent layers of a liquid. It is greater for large molecules than for small. Therefore, the flow of a water system will change when large molecules are dissolved in it. An aqueous system containing hydrophilic starch or soy-protein molecules would be an example of this. This type of viscosity is called the intrinsic viscosity. It is defined as:

$$[\eta] = \lim_{C \to 0} \frac{(\eta - \eta_s)}{C\eta_s} \tag{2}$$

where C is the concentration of the large molecule in grams per milliliter of solvent, η is the measured viscosity and η_s is the viscosity of the solvent itself.

If the molecules under consideration are linear with uniform length and they do not interact with the liquid, Staudinger's viscosity rule[x] can be used:

$$[\eta] = K_m M \tag{3}$$

where K_m is the molecule/liquid constant and M is the molecular weight of the molecule.

If there is interaction between the molecule and the liquid, the simple Staudinger equation does not apply and corrections must be introduced. The initial correction is to raise M to a power from 0.5 to 1.0 to fit the observed data. For the hydrophilic binders used in paper coating colors, this is unacceptable as the interaction is appreciable and hydrodynamic shielding may occur.

If one introduces the fact that such a water-dispersed binder system is really polydisperse and not monodisperse, defining the relationship becomes additionally complex:

$$\eta - \eta_s = \eta_s \sum_i C_i K_m M^a{}_i \tag{4}$$

The situation is further complicated if a hydrophobic additive material is considered. The additive/liquid relationship is predicted by the Einstein equation:

$$\frac{\eta}{\eta_s} = 1 + 2.5\phi \tag{5}$$

where

η	=	the observed viscosity
η_s	=	the viscosity of the liquid
ϕ	=	the volume percent solids.

It should be noted that this equation is designed to be used for solid, monodispersed spheres and is not satisfactory for volume percent solids, ϕ, greater than 10. This, however, is not usually the case.

If the binder system is a mixture of a hydrophilic binder along with a rigid sphere latex, additional complexity is added. The hydrophilic binder relationship (Eq. 4) must be combined with the rigid binder relationship (Eq. 5). In addition to this complexity, the influence of the pigment on the flow properties of the coating color system also must be considered. One could start with the Einstein equation (Eq. 5) and apply corrections for particle shapes other than spheres, polydispersity for each shape, chemical and electrical interactions between each pigment, and interactions between the pigment and the adhesive system. The same consideration would again be necessary for the dispersant, lubricant, insolubilizer, defoamer, or any additional additive.

From this discussion, it is obvious that the complexity of a coating color system will not allow for a mathematical approach in defining the functionality of each and every material in the system, let alone the coating lubricant that is present in a significantly small amount. The best method to understand this complex system is to employ a phenomenological approach. Observe the system and attempt to explain the behavior based upon our limited knowledge of what is present in the system.

Coating flow and Viscosity

The interpretation of the behavior of specific systems requires tools to differentiate and quantify flow behavior. It is necessary to define terminology used for the types of flow that will be observed.

All flow considerations are based upon the response of a liquid to an increase in shear rate. The simplest type of flow is called Newtonian flow or viscosity. Newtonian viscosity is based upon the assumption that no physical or chemical interactions exist between the particles present in the system. Only friction resistance attributed to particle collision is present. The viscosity remains constant as the shear rate changes. Very few systems behave in this manner.

Equation 6 gives a commonly used expansion of the Newtonian relationship which allows other forms of flow to be quantified.

$$\frac{F}{A} = \eta \left(\frac{dv}{dx}\right)^N + f(y) \tag{6}$$

Using this equation for a Newtonian fluid would mean that f(y) = 0 and N = 1.

When a yield value is required to overcome some type of structure in the system, yet the system still exhibits Newtonian behavior, the system is referred to as a Bingham plastic. For this system, N = 1, f(y) is not equal to zero. A Bingham system is idealized and seldom found in practice.

The expression "plastic" is used to imply that a yield value exists for the system, while the expression "pseudoplastic" refers to a system where no yield point exists, but where an increased rate of shear reduces viscosity. Typically, paper coating colors exhibit shear thinning or pseudoplastic behavior. For Eq. 6, pseudoplastic systems show f(y) = 0 and N <1.

If a system that has been subjected to shear exhibits the opposite of pseudoplastic behavior, that is, viscosity increases with increased shear rate, it is said to be dilatent. This type of behavior is due to physical and chemical forces of interaction other than simple friction. For Eq. 6, dilatent systems show f(y) = 0 but N >1. **Figure 7.3** shows the different types of viscosity behavior that can be exhibited by a substance that has been subjected to shear.

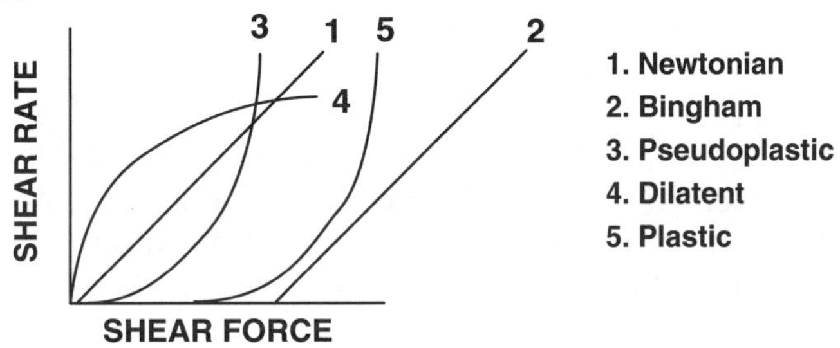

Fig. 7.3. Different types of viscosity behavior

So far, this discussion concerns system behavior that is dependent only upon the rate of shear. When the behavior exhibits a time dependency, the substance has both liquid- and solid-like properties. It is considered to be viscoelastic. "Thixotrophy" is the term used for this time-dependent breakdown of resistance to shear at constant rate of shear. Whereas pseudoplasticity is the breakdown of structure in a system with increased shear, thixotropy is the time-dependent reformation of structure when the shear is decreased. Pseudoplasticity and thixotropy can be realized independently, but this is not typically observed in practice. Since thixotropy is a time-dependent phenomenon, its presence is shown by a "hysteresis" loop. The extent of thixotropy is indicated by the size of the loop. See **Fig. 7.4**. If thixotropy were not present, the structure of the system would reform immediately and an identical response would be exhibited for both increasing and decreasing rates of shear.

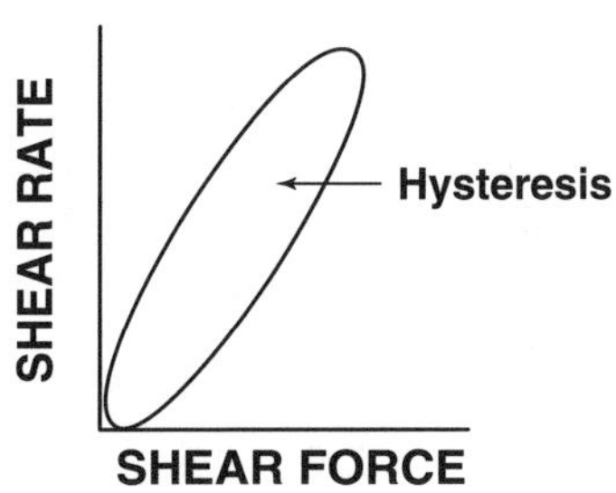

Fig. 7.4. Thixotropy

Wet State Lubrication

A coating color typically exhibits pseudoplastic-thixotropic behavior. Its viscosity will decrease as a function of shear rate, and a time-dependent, structure relationship is present. The magnitude of this wet state viscoelastic character is influenced by the manner in which the various ingredients of the coating color interact with one another. While lubricants are not categorized as true flow modifiers or thickeners, they can impart a profound impact on coating color pseudoplastic-thixotropic behavior. Wet state lubricant performance is believed to be related to the change in flow properties that occur when a metal soap lubricant is present in a paper coating color. Other chemical types of coating lubricants may not appreciably alter coating flow.

Significant low-shear viscosity increases can be realized in coating colors that contain a high portion of a starch binder that has been converted by an oxidative or enzymatic process. This increase in viscosity is usually proportional to the amount of starch and the amount of calcium stearate used in the coating color. Some mills consider this lubricant-related viscosity increase beneficial. In mills where the viscosity change is objectionable, additional starch conversion is performed in order to compensate for it. This can be a problem as starch binding power may be lowered. An explanation for this starch viscosity behavior is based upon the fact that starch has a certain affinity for fatty chemical-type materials. This is demonstrated by an analogous situation whereby the amylose fraction of starch may precipitate when fats and proteins of native starch are present in a starch paste *(16)*.

Heald *(17)* investigated the effect that the addition of a metal salt of a fatty acid has on the viscosity of a starch paste. His objective was to identify a method for determination of the completeness of the hydration of the starch during cooking. This work demonstrated that when relatively high amounts of water-soluble metal soaps are added to a starch cook, a marked increase in viscosity resulted. The magnitude of the increase was proportional to the time of the cook. Under certain circumstances, the entire starch-soap system precipitated as a complex. He concluded that the phenomenon observed was attributed to a specific colloidal interaction between starch and the metal soap. His suggested mechanism was that the starch adsorbed the fatty acid portion of the soap.

The interaction of the lubricant with the hydrophilic binder was used to formulate the lubricant migration theory in coating drying. According to the lubricant migration theory, a lubricant — particularly calcium stearate — migrates to the surface of the coating in a controlled, dispersed form so that it may perform its function at the coated surface *(6)*. The contemporary view of coating application and drying focuses on coating immobilization and pigment–binder interactions. With this view, a migration of a hydrophobic additive through a bound pigment matrix should not occur. Furthermore, the lubricant-plasticizing effect found in the early days of the manufacture of coated paper would not be realized if the lubricant were migrating to the surface of the sheet. While a coating lubricant may be found on the surface of a coated sheet, uniform dispersion in the matrix is a better explanation for its location.

Work examining the compression of polymer monolayers reported that an appreciable and durable surface pressure gradient is formed *(18, 19)*. It was also determined that the introduction of a fatty chemical lubricant facilitated the flow of the monolayer during compression. The lubricant allowed surface pressure gradients to relax more rapidly and more completely. The miscibility of the two components in the film was relevant to the interactions between them *(20)*. The effect also depended upon the amount of lubricant present in the monolayer. Mixed polymer monolayers, in the presence of fatty lubricants, showed greater compressibility when compared to pure polymers *(21)*.

It is interesting to note that even in this simple system, the nature of the dispersion of the lubricant throughout the polymer monolayer was in question. However, based on the effect of the lubricant, some form of dispersion must have occurred. This polymer monolayer discussion has significant impact in the section below that discusses dry state coating properties and the supercalendering process.

The stearate–starch interaction on coating flow properties continues to be of interest to those who manufacture coated grades with starch containing coating colors. It appears to be a function of the type of starch as well as the particle size distribution of the stearate. The phenomenon is not observed when a polyethylene was is used as a lubricant *(6)*. Coating colors containing hydroxyethylated starch binders exhibit the least amount of viscosity increase. Native cornstarch, that has been converted via jet-cooking in the presence of ammonium persulfate, exhibits the greatest amount of viscosity change. **Figure 7.5** shows the difference in Brookfield viscosity as a function of the type of starch for an LWC coating color containing eight parts starch and eight parts SBR latex.

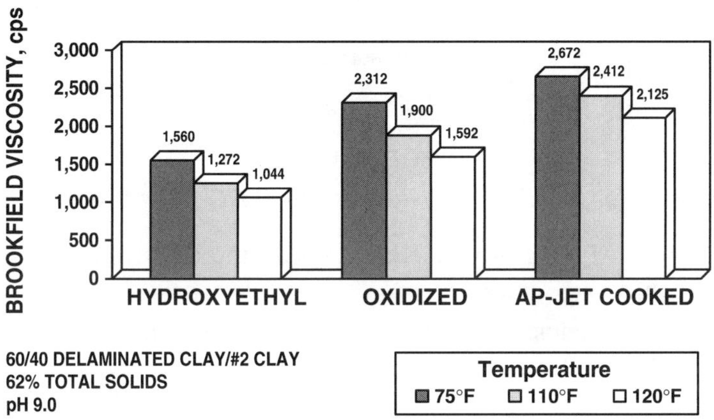

Fig. 7.5. Brookfield viscosity at 100 rpm (LWC: 8 parts starch and 8 parts SBR latex)

Figure 7.6 exhibits the differences in Brookfield viscosity as a function of three different particle size distributions of calcium stearate dispersions. The test was conducted with an LWC coating color containing eight parts of an oxidized starch and eight parts SBR latex. The impact of the calcium stearate particle size distribution on coating color flow properties is a complex issue and should be evaluated for each coating color.

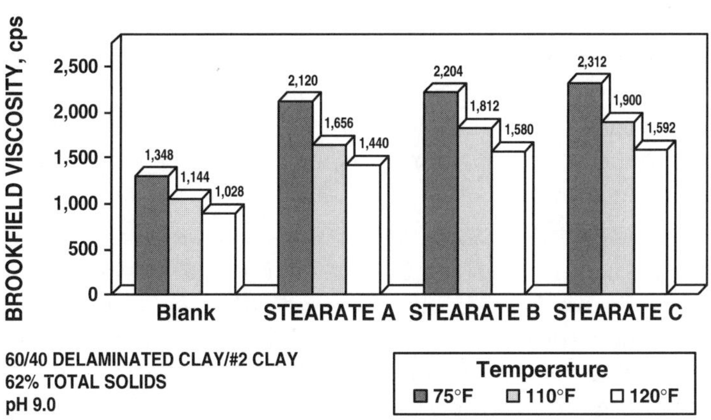

Fig. 7.6. Brookfield viscosity at 100 rpm (LWC: 8 parts oxidized starch and 8 parts SBR latex)

To further demonstrate this viscosity effect, **Fig. 7.7** exhibits an X-Bar control chart for the Brookfield viscosity of an LWC coating color. On the sixth day of this week, a calcium stearate with a relatively fine particle size distribution was introduced into the system. The increase in coating color viscosity was so significant that additional conversion of the starch was necessary to achieve a runnable coating color. Figure 7.2 exhibits the particle size distributions for the two calcium stearate dispersions used.

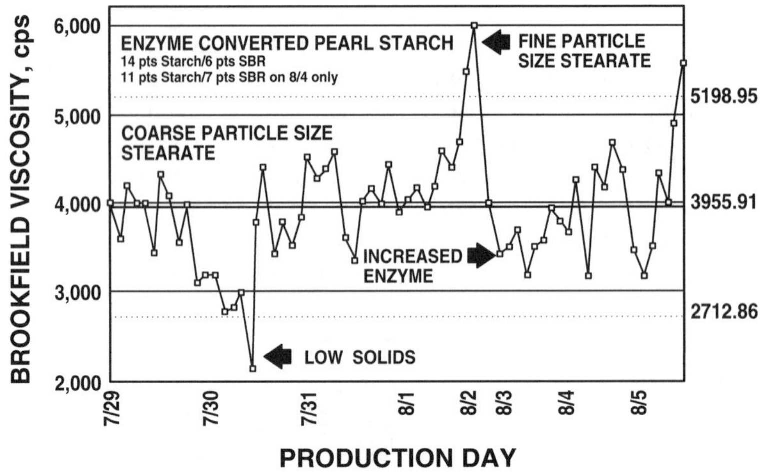

Figure 7.7. X-Bar chart for "trial" Brookfield viscosity (LWC: 40# offset)

Several theories have been advanced to explain the occurrence of viscoelastic phenomena. These include:

1. **Micell Theory**: Particles in a system are aligned by van der Waals Forces. This alignment is affected by lubricants and can be broken by shear stress *(22)*. This is the theory that applies to starch retrogradation and accounts for the pseudoplasticity observed in that system. The starch–soap interactions proposed by Heald *(17)* would be explained by this theory.
2. **Lyosphere Theory**: A particle has a bound shell of the continuous aqueous phase. When the system is subjected to shear stress, this shell is deformed or detached from the colloidal particle *(23)*. It is well known that starch particles, after cooking, have a bound water layer. If the presence of calcium stearate creates a competition for this water or for the starch molecule itself, an increase in pseudoplastic flow could be explained.
3. **Scaffolding Theory**: Mild flocculation occurs which forms a structure within the colloidal system. When shear stress is introduced, the links connecting the structure are broken *(24)*. The lubricant affects the degree of flocculation. This theory works quite well for poorly dispersed pigments and may also fit if a starch–soap interaction exists.

This section reviewed the major types of flow properties that may be experienced when using coating colors. Changes in flow behavior that may be attributable to the use of coating color lubricants have been discussed. It is recommended that those who make coated paper and paperboard critically evaluate coating color formulation changes according to the rheology test procedures found later in this chapter. The coating flow characteristics may be set to meet the needs of the grade being made under the specific manufacturing conditions available. Coating lubricants and all components in the coating color will affect flow behavior.

Dry Coating Lubrication

In providing lubrication at the coated surface, the lubricant performs its most important and best understood function. When a lubricant is present, surface slip characteristics are altered in order to control or eliminate sticking, picking, blocking, and pigment dusting. This permits the papermaker to utilize the minimum amount of binder in order to maintain adequate coating strength. For a better understanding of this lubricant role, it will be necessary to discuss what is known about the science of friction as it may be related to the supercalendering process. An attempt will be made at establishing the contribution that lubricants make to the finishing process.

Tribology and Paper Finishing

Tribology is the science of friction and lubrication and its technical applications. The word tribology originates from the Greek word "tribos" which means the rubbing of surfaces. The essential task of tribology is to identify lubricant needs in order to provide a desired or necessary reduction in friction and wear. Energy savings with cost efficiency during machine operation should be achieved *(25)*.

The first recorded observation of friction was made by Da Vinci in about 1500 A.D. He postulated that friction was independent of the apparent surface area between the sliding surfaces and proportional to the load perpendicular to the surfaces. In 1699, G. Amontons *(26)* identified the fact that friction was material-dependent. The first quantitative law of friction was proposed by Euler. He stated that the coefficient of friction is the ratio of the force applied to the load against frictional resistance to the perpendicular load. The classical laws of friction as stated by Euler are:

1. The coefficient of friction is material-dependent.
2. The coefficient of friction is independent of apparent contact area.
3. The coefficient is independent of load.
4. The static coefficient is higher than the sliding coefficient.
5. The coefficient of friction is independent of sliding speed.

These observations were consistent with what was observed at the time, namely, the use of elastic solid, stone-like objects of high weight that were sliding at low speed over rough terrain under ambient atmospheric pressure. Euler's first law holds true even today. The second law does not hold for highly viscoelastic materials under high pressure. The third law does not apply to viscoelastic materials. The fourth law only applies to low speeds where significant heat cannot be developed to induce phase transitions.

Recently, another type of lubrication phenomenon has gained wide attention, elastohydrodynamic lubrication. Elastohydrodynamics is the study of hard surfaces in contact with elastomers. The major developments in this area are credited to the automotive tire industry.

Elastohydrodynamic Lubrication and the Supercalendering of Paper

The supercalender is a web finishing machine. It is designed to perform an action upon a web of paper such that the surface roughness or nonuniformity is minimized and print smoothness is developed. During the process, the pigment binder will flow, resulting in the orientation of kaolin clay platelets and/or other pigment types with permanent deformation and flow of cellulose and hemicellulose. The type of friction that is of concern in supercalendering is that of grab or traction. It is rolling friction as opposed to sliding friction. The process of supercalendering paper falls into the realm of elastohydrodynamics.

In supercalendering, the elastomeric coated paper surface comes into contact with the steel rolls and the elastomeric or filled calender rolls. Both rolling and sliding friction have coefficients which are defined as the ratio of the applied force to the normal load force:

$$\frac{F}{W} = \mu \tag{7}$$

For elastohydrodynamic friction, the coefficient of friction is a linear sum of contributing components.

$$\mu = \mu_{hysteresis} + \mu_{adhesion} \tag{8}$$

The adhesion component is analogous to the old welding theory, where the points of contact for uneven surfaces between two substances are "stuck" together and require a certain force to pull them apart. In supercalendering, this term would account for the adherence of the paper to the rolls and may be used to explain picking and dusting phenomena. The hysteresis term concerns the friction encountered during deformation of the surface. The rough surface will deform when it is subjected to high pressure, such as the pressure necessary to compress the filled roll of a supercalender. The altered surface character can either elastically rebound to its original position or it can remain permanently deformed. The extent of these processes will be determined by the viscoelasticity of the material. Mathematical expressions can be derived for each component:

$$\mu_{adhesion} = K\left(\frac{\tau_o}{H}\right) \tan\delta = K'\left(\frac{G'}{p}\right) \tan\delta \tag{9}$$

$$\mu_{hysteresis} = K_1 \gamma \left(\frac{K_2}{G'}\right) \tan\delta \tag{10}$$

where:

τ_o = adhesion stress or maximum stress required to break adhesive binding
H = roll hardness
K, K′, K_1 and K_2 = constants of proportionality
tan δ = viscoelastic loss tangent
G′ = shear storage modulus
γ = hysteresis coefficient proportionality constant
p = applied load pressure.

The derivations leading to these expressions are indeed very complex *(27)*. The above equations, although, suggest that supercalendering is a rolling friction process with an adhesion component. This may best explain the paper web's release from the calender rolls during the surface replication process. An influence on pigment removal or dusting may be inferred. The hysteresis component is directly concerned with the supercalendering replication process which results in binder flow and pigment orientation giving print smoothness and surface gloss. It is the shear force that may be found in a rolling nip. In the passage of a paper web through a supercalender, the web receives a push-pull cycle of shear at every nip. In a nine roll stack, there are four filled rolls creating eight finishing nips. The web is subjected to eight plus shear couples alternated with eight minus shear couples or a total of sixteen impulses of shear. This is where finishing takes place. **Figure 7.8** is a drawing of the internal shear strains developed by a filled roll running against a web and a steel roll. Note that the direction of fiber "tilt" implied to the web of paper is the same as inside the filled roll. An analogous situation occurs from the automotive tire industry; namely, stopping a car on short grass would "tilt" the grass blades in the same direction as the tire cords *(28)*.

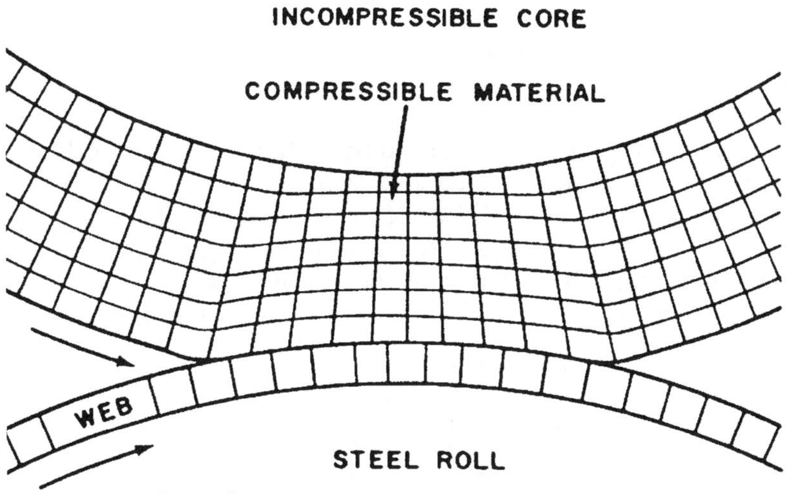

Fig. 7.8. Diagram of the internal shear strains developed by a filled roll running against a web and a steel roll. Exaggerated penetration and nip width *(28)*.

Replication and Lubrication

The desired result from the supercalendering process is the replication of surfaces. This means that the side of the web that faces the steel roll tends to pick up a surface replica of the steel roll surface, while the side that faces the filled roll replicates the filled roll surface. Slippage between either roll and the web surface cannot occur in the medium- to high-pressure areas of the nip, or the replication would be imperfect or smeared. A good analogy to this view of replication is found in the production of matte finished coated and supercalendered paper by using special rough-surfaced or etched steel rolls. These rolls would wear out quickly if slippage existed. While etched rolls, as with any calender rolls, do require refinishing, their life would be only a few hours if slippage occurred.

The supercalendering process, then, is a replication process whereby shear forces are present in a rolling nip. Web slippage does not occur during the pigment orientation as a result of the flow of the binder in the multi-nips of a supercalender. Why, then, is surface lubrication of interest or even necessary?

Surface lubrication requirements arise from a need to maintain or even lower the coefficient of friction comprising the hysteresis and adhesion components. An excellent means for understanding the phenomena of the presence of surface roughness in coated paper may be found by reviewing the work of Skrowronski and Lepoutre *(29)*. Their work demonstrates that two concurrent processes occur during the coating operation: filling up the surface roughness and pores with the coating color and the destructive action of water on the rawstock. The fact that coating rawstock absorbs water and swells during the coating process demonstrates the origin of surface roughness and helps explain why lubricants have a major impact during the supercalendering, slitting, finishing, and subsequent printing processes.

The action of water may result in fiber debonding, internal stress redistribution, and fiber swelling. If these factors act before the blade, they can increase sheet roughness and increase the volume of coating applied; if they act after the blade, they increase the roughness of the coated paper.

The action after the blade is of interest in a discussion of the dry state influence of coating lubricants. See **Fig. 7.9**. That is, while a coated sheet is realized, the irreversible swelling and possible fiber debonding makes for areas that are rich in coating as well as areas that are somewhat deficient in coating. The coat weight that is measured and reported is an average value. The areas that are rich in coating will exhibit significant modulus as the elastomeric coating will easily deform when subjected to the pressures of the supercalender nip. The pigment is sufficiently bonded to the fiber. In terms of elastohydrodynamic friction, this coated area will significantly contribute to the adhesion component of the surface coefficient of friction. In contrasting fashion, the coated areas that are somewhat deficient in coating will offer significant contribution to the hysteresis component of the coefficient of friction. It is desirable to maintain or even lower the total surface coefficient of friction comprised of the hysteresis and adhesion components.

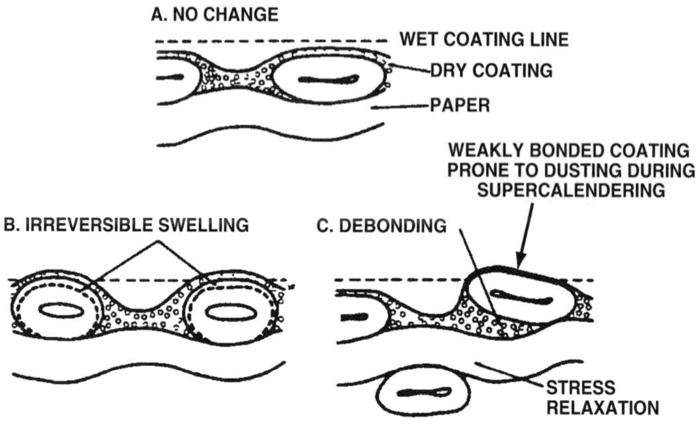

Fig. 7.9. Structural model-change in the surface paper structure: no change (A), irreversible swelling (B) and debonding (C) effects after blade.

From the earlier discussion concerning the lubrication of polymer monolayers *(21)*, one may attempt a mechanism for the lubricant function in the supercalendering process. It was shown in this work that the introduction of a lubricant into polymeric monolayers facilitated the flow of the monolayer during compression and allowed for surface pressure gradients to relax more rapidly and completely. The hysteresis component of the coefficient of friction, then, would concern consolidation of the coating matrix and rawstock as a result of overcoming surface roughness where pressure gradients occur. The presence of the lubricant resulted in a change in the conformation of the polymer molecules at the surface. The low surface pressure achieved through the use of the lubricant strongly suggests that the lubricant molecules penetrate through the polymer strands. This would address the adhesion component of the coefficient of friction. The coating matrix compression or consolidation step would give rise to a migration of the lubricant to the rolls of the supercalender in order to lower the pressure gradients that occur as a result of the surface roughness or nonuniformity.

Those who manufacture coated, supercalendered papers best understand the need for a lubricant in the coating matrix. Attempting to find a definitive mechanism for this additive material requires an investigation in the added complex nature of the dry coating color.

Test methods

All chemical ingredients that are introduced into a coating color have a defined purpose. Before they are introduced into the production cycle, though, their performance should be thoroughly evaluated with laboratory and pilot-scale equipment. Coating lubricants, while difficult to evaluate, are not exempt from this rule.

Wet State Evaluation

For most coating color formulations, coating lubricants influence their viscoelastic properties. Quantification of the magnitude of this influence may be made through the use of several different types of viscometers. Viscometers are available in four different geometries: spindle, bob and cup, plate and cone, and capillary. For those interested in viscosity measurements, the Newtonian relationship (Eq. 1), should be reviewed. The major objective in using any viscometer is to measure the force required to create or sustain a shear rate or sense the shear rate created by any given force *(30)*.

Spindle Viscometers

The workhorse viscometer that the paper industry relies on for low-shear viscosity measurement is the Brookfield viscometer. After the introduction of a coating lubricant into a coating color, a Brookfield viscosity determination should always be obtained and compared with historical data. With the use of calcium stearate, the magnitude of the coating color starch–soap interaction can be most dramatic as measured by Brookfield viscosity. The shear rate for the instrument is based upon the linear velocity of the surface of the spindle, the shape of the spindle, and the distance of separation from the moving surface of the spindle and the stationary fluid. This type of viscometer is used routinely as a process control device by all mills that manufacture coated paper and paperboard. Most coating kitchens exhibit daily displays of coating color Brookfield viscosity values in X-Bar Chart form.

Bob and Cup Viscometers

Traditionally, the Hercules high-shear rheometer has been the only bob and cup type viscometer that has been used in the evaluation of coating colors. Other manufacturers, such as Haake, also supply this type of viscometer. The latest of the Hercules models offers computer control with strain gauge accuracy. The type of shear is annular, and the rate is controlled by the accuracy of the distance between the bob and the cup. A wide range of shear rates are possible, up to 10^3 sec^{-1}.

With the advent of the computer controlled Hercules high-shear rheometer, it is now possible to determine the various components of the rheological evaluation. Typically, Hercules rheograms are evaluated through comparison practices. Even in the comparison mode, it is a rather easy task to quantify coating color rheological behavior.

Cone and Plate Viscometers

For cone and plate viscometers, changing the small angles of the cone allows for significant variation in the shear rates that may be achieved. A relatively small amount of sample is required. High-shear rates may be achieved by using extremely small cone angles, but care must be taken as edge effects must be considered so that the point of the cone is always in contact with the plate. All cone and plate viscometers are computer- and temperature-controlled. All components of the rheological profile may be determined.

Capillary Viscometers

For capillary viscometers, the shear rate is a function of the radius and flow rate through the capillary. With this type of instrument, shear rates as high as 10^6 to 10^7 sec^{-1} are possible. Instruments that utilize gravity and pressure flow are available. All the instruments available are computer-controlled so that a total rheological profile is possible.

Pilot Paper Coating Trials

While the rheological profile quantification capabilities are significant through the use of sophisticated viscometry, high-speed pilot-scale coating trials have the added advantage of introducing the rawstock variable into the lubricant evaluation process. Additionally, dry state coating color properties may also be evaluated. Sensor and Simulation Products' Cylindrical Laboratory Coater is an ideal tool for pilot paper coating trials involving coating lubricants.

Dry State Evaluation

In the dry state coating matrix, lubricants are used in order to alter the surface and/or interfacial tension of the coating color. For glossy grades, pilot-scale supercalendering or coated sheet samples should be performed under temperature and pressure conditions typical to the mill conditions. If dusting or sticking is normally a problem at high pressures or temperatures, the function of the coating lubricant can usually be observed during this evaluation.

Tests for the Finished Sheet

Antidusting

While the problem of poorly bound pigment can best be evaluated during supercalendering, some undesirable dust may remain with the coated sheet. There are several tests for the presence of dust on the surface of a coated sheet. All are subjective in nature and involve abrading the coated sheet by rubbing a black velour paper against the surface. The quantification of the pigment present on this black paper is done in a comparative manner. Test methods may be found in the TAPPI PRESS report, "Additive Testing and Paper Evaluation."

Sheet Gloss

Sheet gloss should be measured according to TAPPI Test method T 480 om-92, "Specular gloss of paper and paperboard at 75 degrees." Both Hunter and Technidyne manufacture instruments that meet the requirements of this TAPPI Test method.

Print Quality and Print Gloss

Print quality and the quantification of any printing test is somewhat subjective. Proofing presses that simulate the actual press room conditions are best for this evaluation. For pilot-scale testing an IGT, PrufBau or similar print testing equipment should be used. These printing devices offer test methods that may be unique to the instrument.

Wet Pick and Wet Rub

The best wet pick test is a lengthy press run on an offset press. For laboratory evaluations, the test methods for the IGT print tester fitted with the wet pick test apparatus as well as the PrufBau print tester should serve well in the evaluation of wet surface characterization. Another instrument that is useful for wet rub testing is the Adams tester which provides a static testing of water resistance of coated paper and paperboard.

Traditional Dry Tests

According to the desire and needs of the particular evaluation, traditional paper testing should be performed according to the appropriate TAPPI Test method *(31)*. As lubricants may influence the viscoelastic character of coating colors, such optical properties as brightness and opacity should be determined. Smoothness, both Sheffield and Parker PrintSurf, will help quantify the efficiency of the supercalendering step.

Literature Cited

1. Baumeister, M. and Kraft, K., *Tappi* 64(1): 85(1981).
2. Lepoutre, P., *Paper Trade J.* 72(9): 87(1984).
3. Skowronski, J. and Lepoutre, P., *Tappi J.* 68(11): 98(1985).
4. Beck, U., Goossens, J.W.S., Rahlwes, D. and Wallpott, G., *Wochbl. Papierfabr.* 108(13): 499(1980).
5. Bundy, W. M., Harrison, J. L., and Ishley, J. N., *Tappi J.* 66(): (1983).
6. Hern, J. F., *Paper coating additives,* TAPPI PRESS, Atlanta, 1978, pp. 80–105.
7. Henkel Emery Group, "Commercial Fatty Acids," Cincinnati, 1991, p. 8.
8. U.S. Patent 2,425,828, Retzsch and Lighthipe, NOPCO Chemical Company, Dec. 12, 1944.
9. Allied Chemical Company, "Polyethylene Waxes for Industry," Morristown, NJ, 1986, p. 4.
10. U.S. Patent 4,659,489, Hill and Flynn, Diamond Shamrock Chemical Company, April 21, 1987.
11. U.S. Patent 4,676,836, Hill and Flynn, Diamond Shamrock Chemical Company, June 30, 1987.
12. U.S. Patent 4,997,479, Hou and Hui, P. P. G. Industries, Inc., March 5, 1991.
13. U.S. Patent 4,766,015, Nikoloff and Peltier, Bercen Inc., August 23, 1988.
14. Adams, A., TAPPI PRESS Report No. 1349, "Paper coating additives, Description of Functional Properties and List of Available Products and Suppliers," May 1990.
15. Doherty, D. J., *Science of Surface Coatings,* Van Norstrand, W. Chatfield, 1962, p. 365.
16. Voigt, J. and Krasniewski, J., *Starch and Starch Products in Paper Coating,* TAPPI PRESS, 1990, p. 71–91.
17. Heald, A. M., *Paper Trade J.* 29(7): 21(1941).
18. Peng, J. B. and Barnes, G. T., *Langmuir,* 1990, 6, 578.
19. Peng, J. B. and Barnes, G. T., *Langmuir,* 1991, 7, 1749.
20. Gaines, G. L., *Insoluble Monolayers at Liquid-Gas Interfaces,* Interscience Publishers, New York, 1966, Chap. 5.
21. Peng, J. B. and Barnes, G. T., *Langmuir,* 1991, 7, 3090.
22. Alexander, J., *Colloid Chemistry,* Vol VI, 1946, p. 342.
23. Goodeve, C. F., *Trans. Faraday, Soc.* 35:342(1939).
24. Eyring, H., *J. Chem. Phys.* 4: 283(1936).
25. Klamann, D., *Lubricants and Related Products,* Verlag Chemie GmbH, Weinheim, Germany, 1984.
26. Amontons, G., *De la restance Coussie dans les machines,* Histoire Acad. Roy. Sci., Paris 12, 200 (1699).
27. More, D.F., *The Friction and Lubrication of Elastomers,* Pergamon Press, 1972.
28. Pfeiffer, J.D., *The Finishing Action of A Supercalender Nip,* Beloit Corporation, 1969.
29. Skowronski, J. and Lepoutre, P., Pulp and Paper Research Institute of Canada Report No. 507, December 1984.
30. Kline, J. and Diehm, R., *Synthetic Binders in Paper coatings,* TAPPI PRESS, 1975, p. 4–21.
31. TAPPI Test Methods:
    T 423 om-89 "Folding endurance of paper"
    T 425 om-91 "Opacity of papers"
    T 458 om-89 "Surface wettability of paper"
    T 474 wd-71 "Bending quality of paperboard"
    T 477 m-47 "Blocking resistance of paper and flexible materials"
    T 479 om-91 "Smoothness of printing paper"
    T 480 om-92 "Specular gloss of paper and paperboard at 75 degrees"

T 499 su-64 “Surface strength of paper”
T 503 om-94 “Coefficient of static friction of shipping sack papers”
T 511 om-88 “Folding endurance of paper”
T 514 cm-92 “Surface strength of coated paperboard”

CHAPTER 8: INSOLUBILIZERS

William C. Floyd and Susan F. Boss

Introduction

Paper coating insolubilizers are primarily used to reduce the water solubility or sensitivity of the pigment binders used in the coating "colors" or formulations. Water resistance is important for many applications of coated paper or paperboard in which the coated substrate may somehow be moistened. Insolubilizers may also strengthen the binder and thereby improve the surface strength of the coated sheet. In many printing applications, a coated sheet requires wet-rub and wet-pick resistance because of aqueous fountain solutions and the mechanical stresses of the printing process. The myriad of applications employed for coated paper and paperboard requires different types of binders and different types of insolubilizers because they have different requirements for water resistance. Some binders, such as starch, are inherently water-sensitive and are more difficult to insolubilize. Others, such as a synthetic latex, are inherently more water-resistant and have a lesser need for insolubilization.

In recent years, the type of insolubilizer has been shown to affect the printability or print quality of coated substrates. The insolubilizer can affect the receptivity of the coated surface to the oil and water used in the printing process. This receptivity influences such qualities as dot definition and print clarity.

Terminology

The term "water resistance" is of central importance in any discussion of coating insolubilizers. Water-resistant must be distinguished from "waterproof." The latter term refers to a substance which has been rendered completely impervious to water by a hydrophobic material such as wax. Water resistance may be broadly defined as an impedance of the adverse effects of exposure to water. Most printing paper coatings are somewhat porous and easily wetted by water. Penetration of water is acceptable, but the critical factor is the degree to which the binder is softened or solubilized. Softened binders result in loosened pigment particles. The water resistance may be evaluated by various wet-rub tests. These tests are often used to gauge the efficiency of an insolubilizer. In addition to water resistance, the following definitions may be useful in discussing the conditions affected by specific paper coating additives. These terms were also included in the 1978 revised edition of the monograph *(2)*.

- **Dry Pick Improvement** — To increase the resistance to damage or removal of the dry coated surface when subjected to the pulling stresses encountered on an offset printing press.
- **Dry Rub Improvement** — To increase the resistance to damage or removal of the dry coated surface when frictional forces are applied by dry adjacent surfaces.
- **Wet Pick Improvement** — To increase the resistance to separation or removal of the coated surface from the base sheet when subjected to the pulling stresses encountered on an offset printing press, wherein either or both surfaces may be wet or moist due to water pickup by rewetting or due to high humidities.
- **Wet Rub Improvement** — To increase the resistance to damage or removal of the wetted or moistened coated surface when frictional forces are applied by wet adjacent surfaces.
- **Water Resistance** — To reduce the rate of penetration or absorption of water into the coated surface.

Another term that is useful when reviewing insolubilizer technology is "piling." This term comes from the printing industry and describes one of the major problems associated with the adverse effects of water during offset printing. One reference defines piling as the accumulation of paper coating on the blanket of the offset press, thereby interfering with proper ink transfer *(143)*.

General References

The earlier technology of paper coating insolubilizers has been well covered by Ritson, Greif, and Stonebraker in TAPPI Monograph No. 25, "Paper coating additives" *(1)*. The coverage of this technology was updated by Landes, Kroll, and Ritson in the revised edition of this monograph *(2)*. Their revision covered the period through the mid-1970s and served as an important resource for this chapter. Much general information is available from texts by earlier authors such as Nadelman and Baldauf *(3)*, Loomer *(4)*, Casey *(5)*, Gunning *(6)*, and Busch *(7)*. A number of other TAPPI monographs cover topics relating to the use of insolubilizers, and some have been or are being updated to include current information *(8–10, 17, 36, 37)*. Some references primarily directed toward the printing industry may also be of interest in the formulation and testing of paper coatings *(143, 144)*.

Factors Affecting Insolubilizer Developments

In the late 1970s and early 1980s, starch was the predominant natural binder used in paper and board coatings. Protein, polyvinyl alcohol, and casein were used to a lesser extent. These binders were generally used with various amounts and varieties of synthetic latices. The particular formulation depended upon the grade of paper and the end-use application.

These natural binders required insolubilizers, and melamine-formaldehyde or urea-formaldehyde resin was used most often. To a lesser extent formaldehyde, glyoxal, and certain metal salts were used, but problems associated with these products limited their usage. A major factor behind a change in insolubilizer technology was a strong desire to minimize or eliminate formaldehyde exposure in the workplace. This was prompted by the publishing of several controversial reports linking formaldehyde exposure to health risks *(11–13)*. Environmental concerns effected some changes in the paper industry. Increasing energy costs prompted coating solids to be increased to improve drying efficiency. As paper technology advanced, coating speeds increased. A trend toward alkaline paper-making emerged. As the paper market became more competitive, quality improvements were demanded. Printability and performance on the printing press became even more important.

It is important to recognize that the degree of water resistance imparted or the wet-rub or wet-pick requirements of a particular grade of paper are not the sole factors for determining the choice of insolubilizer. The type of binder is very important, as are pH limitations, viscosity responses, surface receptivity, printability, end-use applications, and economics. Care should be taken to optimize the level of insolubilizer. The fact that a low level works well does not necessarily mean that more will work better. Often excessive levels of insolubilizer may result in increased water sensitivity, brittleness, or fold cracking.

Mechanisms and Methods

Ritson divided the various known methods of improving the wet-rub and wet-pick performance of paper coatings into three main classes *(14)*. The recent developments in insolubilizer technology may also be fitted into this grouping.

1. Formation through chemical reactions of coatings that are water-insoluble or have reduced water sensitivity. Examples are:
   a. Metal salts of calcium, zinc, aluminum, chromium, antimony, or zirconium that react to form insoluble products or complexes with reduced water sensitivity.
   b. Aldehydes or aldehyde-based derivatives that react with the functional groups (such as amino or hydroxyl) to produce crosslinking with reduced water sensitivity.
   c. Aminoplast resins that react with amino or hydroxyl groups of the binder to produce crosslinking or blocking of these groups with reduction in water sensitivity.
2. Incorporation of binders of low water sensitivity to protect water-soluble types of binders. Examples are:
   a. Emulsion polymers or latices, such as some of the vinylacetate, acrylic ester, or styrene-butadiene types have been used as cobinders with improvements in water resistance.
   b. Water-soluble polymers (such as polyvinyl alcohol) which develop some water insolubility upon drying can produce improved water resistance when blended with the more water-sensitive starches.
3. Addition of highly water-repellent materials such as wax sizes or metallic soaps. Used in small proportions these may provide some water resistance but do not give effective wet-rub resistance after water has penetrated the coating.

Methods of Application

There are several methods by which insolubilizing agents have been introduced into paper coatings:

1. The most common method is to add the insolubilizer as a solution or dispersion to the coating color either during or after color preparation, but before application of the coating color to the sheet. It is generally preferred to add the insolubilizer as the last ingredient of the coating color to minimize premature reaction with the binder.

2. A method often used for paperboard is to apply a wash coat of insolubilizer in dilute solution to the surface of the side of the sheet which will be coated by a calender water box. Other means of application such as sprays, squeeze rolls, or immersion may also be used.
3. A base sheet is treated with a solution or base coating containing the insolubilizer (often in excess), and then a top coat containing the binder to be insolubilized is applied. By this procedure, it is sometimes possible to avoid premature thickening that can result from reactions between certain insolubilizers and binders.
4. A gaseous form of the insolubilizer (such as formaldehyde) may be applied to the coated sheet as it passes through a closed chamber. This method has little if any commercial application at this time.

Obviously the chemical and physical nature of the insolubilizer and its interactions with the binder and other coating components must be considered when deciding upon the most practical method of application.

Types of Insolubilizers

Many chemical types of insolubilizers have been proposed and tried in pigmented paper coatings. The literature on insolubilizing various binder types over the years is voluminous. Many of the insolubilizers mentioned in the literature are not currently used commercially in paper coatings because of high cost, poor availability, toxicity, efficiency, or quality deficiencies.

The insolubilizers used by paper coaters on a commercial basis include the following types of products (excluding any polymers that might also be classified as coating binders.):

1. Formaldehyde (various forms)
2. Glyoxal
3. Aminoplast resins
4. Modified glyoxal or glyoxal-based resins
5. Metallic salts
6. Other various types.

A detailed discussion of these types of insolubilizers follows, including their background, mode of action, and usage.

Formaldehyde

Historical Perspective

Formaldehyde (HCHO) *(9, 15, 16)* is the simplest member of the aldehyde family. Formaldehyde is commonly available in the form of aqueous solutions containing 37–50% HCHO by weight. These aqueous solutions may be stabilized by various amounts of methanol. Formaldehyde is also available as paraformaldehyde, a solid hydrated polymer, and as trioxane, a cyclic trimer. Hexamethylene tetramine is produced from formaldehyde and ammonia and is used as a source of formaldehyde in certain applications. An excellent reference on formaldehyde is the monograph by Walker *(16)*.

Formaldehyde was one of the earliest materials to be considered as a crosslinking agent for starch in various applications *(17–19)*. Formaldehyde requires a low pH to fully crosslink starch, and this condition is not compatible with most coating formulations. For this reason, formaldehyde did not experience significant usage as a starch insolubilizer.

However, formaldehyde has been found to be very effective as a protein insolubilizer *(9)*. Formaldehyde can be used to impart water resistance to such proteinaceous substances as animal glue, soy protein, and casein. Formaldehyde was once the dominant agent for insolubilizing protein in paper coatings, but has lost ground in recent years to other technologies. Formaldehyde is unpleasant to work with because of its irritating properties. It has also reportedly been associated with health hazards *(11, 12)*. Gloves, goggles, a respirator, and other protective gear should be employed when handling formaldehyde.

Chemistry

Formaldehyde in aqueous solution exists as a hydrate, methylene glycol, in equilibrium with the polymer polyoxymethylene $(HO(-CH_2-O)-_nH$ *(16)*. The initial reaction between formaldehyde and a starch hydroxyl group forms a labile hemiacetal. The hemiacetal is most stable in the pH range of 5–7. Full crosslinking results from acetal formation at a lower pH *(18)*. This reaction is depicted in **Scheme 8.1.**

Polyvinyl alcohol and formaldehyde would follow a similar course.

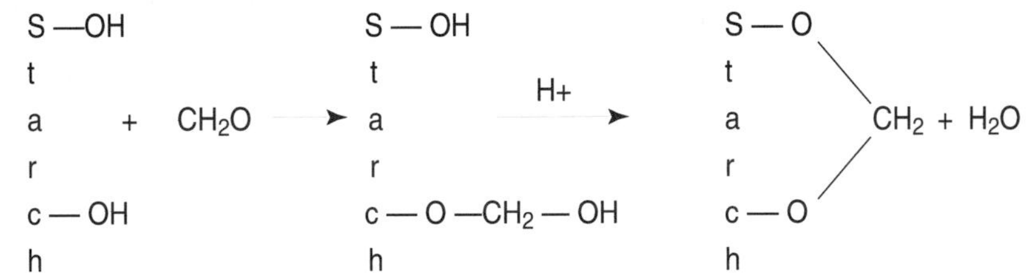

Scheme 8.1. Reaction of formaldehyde with starch

Proteins are quite complex in structure. They comprise polypeptide chains built up from amino acid residues. These contain both amino and carboxylic acid groups as well as hydroxyl and sulfhydryl groups *(9, 20)*. Formaldehyde reacts in various ways with protein, but the main reactions leading to insolubilization are seen in **Scheme 8.2** *(9, 20)*.

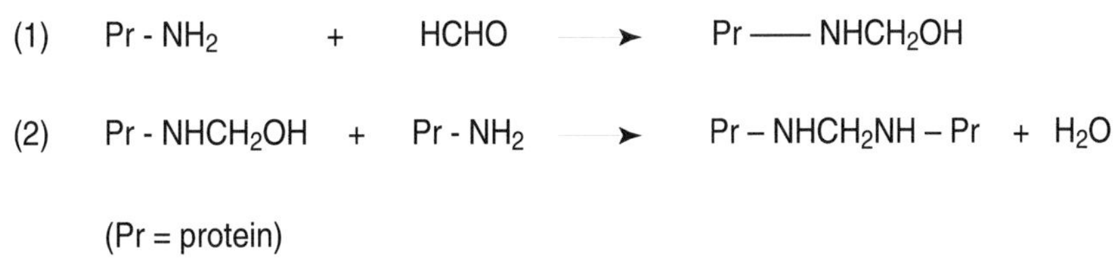

Scheme 8.2. Reaction of formaldehyde with protein

The amino functionality of the protein reacts with the formaldehyde under alkaline conditions to form a methylol compound. This methylol compound then reacts with another amino group, expelling water, to form a crosslink. These reactions are desirable only after the coating has been applied. The heat required to dry the coated paper is often not sufficient to develop the best wet-rub or wet-pick values. Some additional heat in a curing stage may be required for a complete cure. Wet-pick resistance may require more time to develop than wet-rub resistance, but generally follows the same trend *(20)*.

Current Status

Formaldehyde is not used commercially to insolubilize starch or polyvinyl alcohol. Formaldehyde in various forms is used to insolubilize protein, but this is believed to be of minor importance. Use of formaldehyde and worker exposure to formaldehyde are covered under U.S. government regulations *(21)*. Most of the formaldehyde used is added as a dilute solution with ammonia to the coating just prior to application. In many protein coating applications, formaldehyde has been replaced by melamine resins, blocked glyoxal resins, or certain metal salts. Detailed recommendations on the use of formaldehyde in protein coatings are provided in TAPPI Monograph No. 36 *(9)* or may be obtained from a protein supplier.

Glyoxal

Historical Prospective

Although glyoxal has been known for a long time, it was still considered a "specialty" chemical in the 1950s and did not become a true industrial product until the early 1960s *(22)*. The early products contained impurities that contributed to color problems, and some contained considerable amounts of formaldehyde and acidic byproducts. The quality of the commercially available glyoxal has improved greatly in recent years, and any impurities or by-products are now at trace levels. Glyoxal has been used with some success to insolubilize natural binders. Many references describe its use with protein *(9, 23)*, starch *(8, 23, 24)*, and polyvinyl alcohol *(10, 23, 25, 26)*. The high reactivity of glyoxal often presents viscosity problems. The pH limitations and color formation may limit its usefulness in protein formulations.

Chemistry

Glyoxal is the simplest aliphatic dialdehyde. Anhydrous monomeric glyoxal is unstable and rarely isolated *(22, 27, 28)*. The most common form of glyoxal is the commercially available 40% aqueous solution. In aqueous solution, glyoxal exists in the form of a completely hydrated monomer (**Fig. 8.1**), dimers, trimers, and other oligomers (**Fig. 8.2**). This structural information was determined by proton and carbon-13 nuclear magnetic resonance experiments *(29–31)*. These studies supported the 5-membered dioxolane structures shown and demonstrated the absence of the 6-membered dioxane-type structure previously accepted. The amount of each specie in solution is concentration dependent. Aqueous glyoxal solutions have low vapor pressure and little odor.

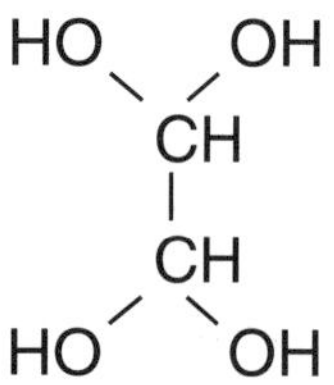

Fig. 8.1. Hydrated glyoxal monomer

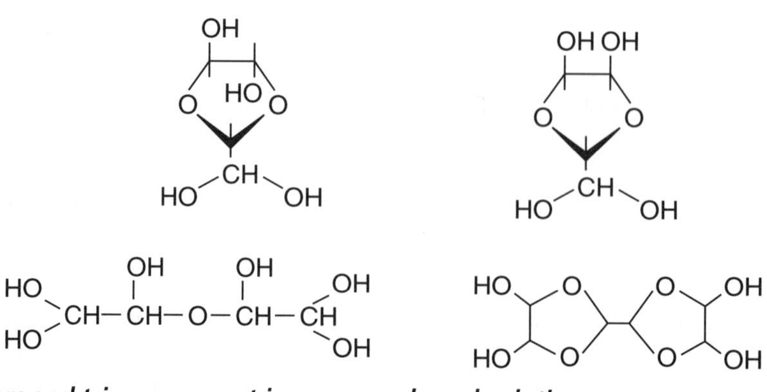

Fig. 8.2. Dimers and trimers present inaqueous glyoxal solution

Glyoxal in aqueous solution reacts rapidly with hydroxide ion (HO^-) undergoing an internal Cannizzaro reaction to form glycolate ion *(32, 33)*. (See **Scheme 8.3**.) This reaction is minimized at low pH and is why commercial glyoxal samples are received at a pH of 1 to 3. The rate of the Cannizzaro reaction increases rapidly as pH increases into the alkaline region. The use of glyoxal in highly alkaline coating formulations is not recommended. Glyoxal undergoes many reactions typical of aldehydes such as formation of hemiacetals and acetals with hydroxylated substrates such as starch and polyvinyl alcohol. Glyoxal also reacts with amino groups and may be used to crosslink protein.

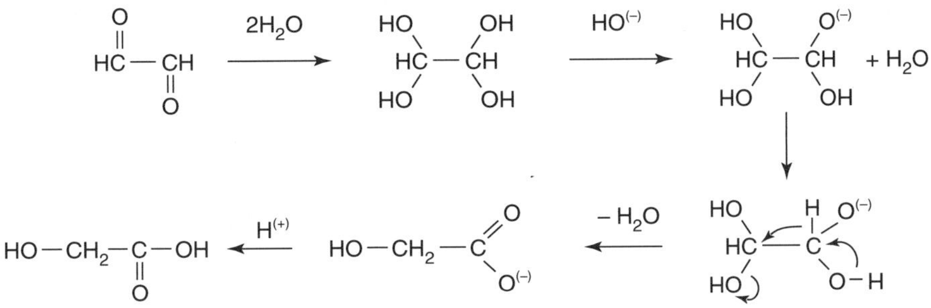

Scheme 8.3. Cannizzaro reaction of glyoxal

Use of Glyoxal

Much additional technical information with references on glyoxal is available in technical bulletins from commercial suppliers *(22, 27, 28)*. Glyoxal has been used to impart temporary wet strength to paper *(22, 27)* and as a crosslinking agent to impart wet-rub resistance to pigmented coatings. Since the 1960s, several articles provided extensive information on the use of glyoxal as a binder insolubilizer. Buttrick and Eldred *(24, 26)* covered its use in starch-pigment coatings. Buttrick *et al. (23)* expanded this work with starch to cover the insolubilization of coating binders based on casein, soy protein, and polyvinyl alcohol. TAPPI Monograph No. 17 *(8)* contains additional information and references on the use of glyoxal in starch-based coatings. There are many important variables that affect the use of glyoxal as an insolubilizer including type of binder, amount of glyoxal used, coating color solids, pH, and other coating additives. The use of aqueous glyoxal as an insolubilizer has diminished in recent years with the advent of blocked glyoxal resins.

With Protein

Glyoxal reacts rapidly with the amino functionality of protein as shown in **Scheme 8.4** *(20)*. This crosslinking reaction can be used to provide suitable wet-rub resistance to coated papers containing protein-based binders. The unsaturated crosslink resulting from dehydration of the adduct is believed to give rise to the development of undesired color formation. This color development can contribute to a loss of brightness.

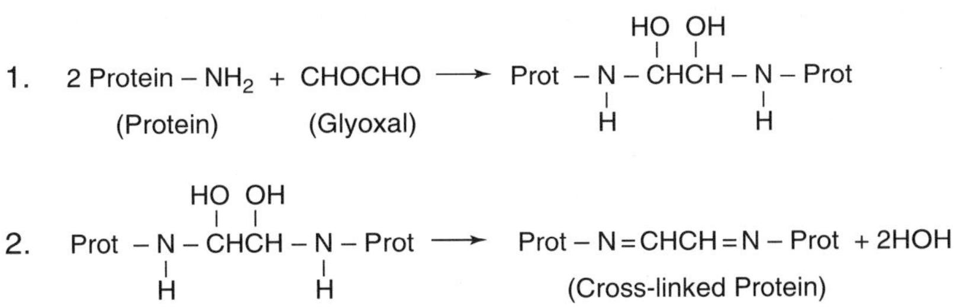

Scheme 8.4. Reaction of protein with glyoxal

Typical laboratory results using glyoxal to insolubilize casein-clay and soy protein-clay coatings were reported by Buttrick *et al.* *(23)* and are shown in part in **Table 8.1**. Acceptable wet-rub resistance was attained with 3–5% active glyoxal on the weight of the protein. Adverse effects noted were the increase in viscosity and a drop in brightness.

Table 8.1. Properties of glyoxal — casein and glyoxal — soy protein coatings

	Insolubilizer, %[a]	Coating color Visc., cp[b]	Brightness %	Wet Rub Resis.[c]
Glyoxal-Casein				
	0	3680	79	10
	1	7870	78	23
	3	7740	78	86
	5	9000	78	98
Glyoxal-Soy Protein				
	0	1470	79	4
	1	1910	78	18
	3	1850	76	89
	5	2010	76	97

[a] *Insolubilizer (active basis) on weight of protein*

[b] *Brookfield viscosity, no. 4 spindle, 60 rpm, LV viscometer.*

[c] *Wet rub resistance, modified TAPPI Useful Method 462. High readings show best rub resistance.*

Glyoxal has had a limited use in commercial protein-based coatings. The color development and viscosity problems may be partly responsible for this limited use. The most critical factors limiting the use of glyoxal with protein may be the high alkalinity and pH required to solubilize the protein. At a pH above 9.0, glyoxal is soon destroyed. Excessive amounts of ammonia can divert the reaction of glyoxal from protein to the formation of imidazoles.

With Polyvinyl Alcohol

Although polyvinyl alcohol provides somewhat more water resistance than starch when used as a binder in paper coatings, it is generally not satisfactory for use in applications requiring a high degree of water resistance without the use of insolubilizers.

Sufficient glyoxal solution to give 10% active ingredient, based on polyvinyl alcohol has been recommended to insolubilize it *(34)*. Higher dialdehydes such as glutaraldehyde can also be used to improve the water resistance of polyvinyl alcohol *(34)*. The reaction is reported to be catalyzed by acidic materials and accelerated by heating.

With Starch

The use of glyoxal as an insolubilizer for starch binders in paper coating has been discussed in several references *(8, 23, 24)*. A major factor in the water resistance development of glyoxal-insolubilizer starch coatings is the type of starch employed *(35–37)*. The film-forming ability and binding strength of the starch is indicative of the relative degree of water resistance that might be anticipated with various starches and a given amount of insolubilizer.

In general the viscosity of a starch-based coating color will increase with increasing glyoxal concentration, with increasing starch concentration, or with increasing coating color solids. Glyoxal causes a greater viscosity increase with high molecular weight (high viscosity, i.e., low fluidity) starches than with low molecular weight starches. This is true regardless of whether the starch is acid-modified, enzyme-converted, oxidized, or hydroxyethylated. In low-solids coatings, such as those employed for a sizepress or air knife coating, the increase in viscosity from glyoxal is usually negligible. In high-solids coatings, such as those employed for a blade or rod coater, the increase in viscosity from glyoxal is more pronounced. Significant quantities of glyoxal are used in some European and Scandinavian countries, often for offset printing applications *(142)*. These formulations generally employ highly converted starches (having low viscosity) which minimizes adverse rheological properties.

Glyoxal is reported to undergo three possible reactions with starch *(142)*. These are shown in **Scheme 8.5**. Reaction (1) forms a reactive mono adduct of glyoxal with the hydroxylated substrate such as starch or PVOH. This intermediate may further react with an adjacent hydroxyl as in (2) or with distal hydroxals, as in (3) to form a crosslink.

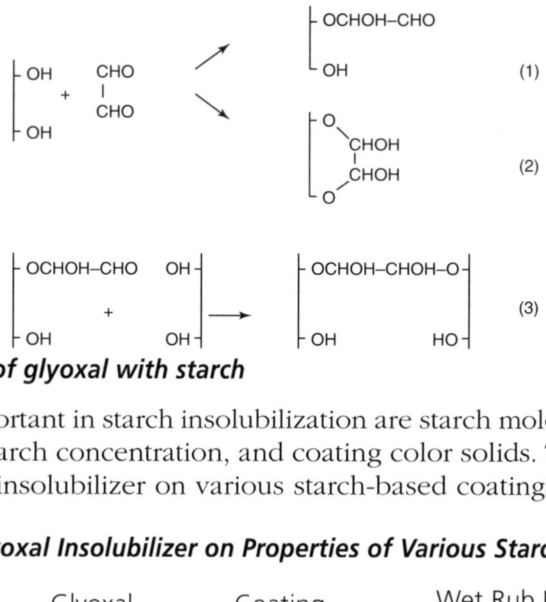

Scheme 8.5. Reaction of glyoxal with starch

Other factors important in starch insolubilization are starch molecular weight, glyoxal concentration, pH, starch concentration, and coating color solids. **Table 8.2** *(23)* shows the effect of glyoxal insolubilizer on various starch-based coatings.

Table 8.2. Effect of Glyoxal Insolubilizer on Properties of Various Starch Coatings

Type of Starch (Corn)	Glyoxal %[a]	Coating Visc. cp[b]	Wet Rub Resistance, %[c]	
			Air Dried	Cured[d]
Hydroxyethyl	0	2320	5	8
Hydroxyethyl	2.5	3980	83	88
Hydroxyethyl	5.0	5560	91	97
Oxidized	0	3340	3	6
Oxidized	2.5	4720	26	57
Oxidized	5.0	5840	54	73
Enzyme Converted	0	1200	2	4
Enzyme Converted	2.5	1400	35	75
Enzyme Converted	5.0	2400	55	95

[a]*Glyoxal (active), % on dry starch*

[b]*Brookfield viscosity, no. 3 spindle, 10 rpm*

[c]*Wet rub resistance, modified TAPPI Useful Method 462. High readings show best rub resistance.*

[d]*Curing conditions — 10 min at 110°C*

Advantages noted for using glyoxal as an insolubilizer have included: rapid cure at low temperatures (often having full wet-rub resistance as soon as the sheet is dried), lack of formaldehyde fumes, and ease of storage and handling of glyoxal solutions. Disadvantages noted have been: sensitivity to high pH (over 9), thickening, short pot life, rheology problems, color development, and limitations on the use of highly alkaline pigments. Viscosity increases are reduced by addition of latex to the coating formulation. However, some highly carboxylated, high pH latices may not be compatible with glyoxal. The use of starch viscosity modifiers, such as urea *(41)* or dicyandiamide *(42)*, are not recommended with glyoxal. These products react rapidly with the glyoxal *(40)* and reduce its efficiency as an insolubilizer. For optimum efficiency of glyoxal in starch-based coating colors, the pH should be maintained in the 7.0–7.5 range.

Two-Coat Applications

Coated boards for offset and gravure printing are frequently double-coated using a natural cobinder and latex in the base coat and latex in the top coat. Insolubilizing agents are usually needed in both coatings to promote adhesion between the two coatings and to impart good water resistance to the top coating. An insolubilizing agent, such as glyoxal, is incorporated in a starch base coat and applied to board, followed by application of the desired top coating. The insolubilizing agent should not react materially with the base coat binder during application *(38, 39)*. Oxidized and dextrin-type starches are particularly suited for this application. After both coatings are applied, the insolubilizing agent should diffuse or migrate to the top coating and insolubilize the natural binder.

The following formulations in **Table 8.3** demonstrate the use of glyoxal in a base coat application. Although a trailing blade base coat is shown, roll or sizepress coaters can also be used to apply glyoxal-starch base coats.

Table 8.3. Two-coat applications for offset board

Starch base coat — trailing blade coater

Ingredients	Parts by Weight (Solids)
No. 2 clay	70
Ground calcium carbonate	30
Natural binder	5
Latex	15
Glyoxal	1
Lubricant	1
Total solids	60–65%
pH	8.0
Brookfield viscosity (cps @ 100 RPM)	1500

Protein-Latex Air Knife Top Coat

Ingredients	Parts by Weight (Solids)
No. 1 high brightness clay	85
No. 2 clay	15
Low viscosity protein	5
Latex	15
Lubricant	1
Total solids	45%
pH	9.0
Brookfield viscosity (cps @ 100 RPM)	90

Glyoxal is used today, to a limited extent, in starch and starch-latex coating formulations. However, its greatest use in coatings is in the form of blocked glyoxal resins, which will be discussed later. These resins combine the benefits associated with the use of glyoxal as an insolubilizer and overcome many of the disadvantages.

Amino Resins

Historical Perspective

Amino resins is the term commonly applied to the condensation products of aldehydes and polyfunctional amides and amidines. Formaldehyde is, by far, the most commonly used aldehyde, with urea-formaldehyde and melamine-formaldehyde resins being the most widely used amino resins in paper coatings. In most cases, the amino-functional substrate is reacted under alkaline conditions with the formaldehyde to form a reactive monomer. This monomer may be alkylated for stability or carefully condensed to give various types of polymeric intermediates. These products are heat-sensitive and, with acid catalysis, can be further polymerized into hard, heat-resistant forms. These intermediate resins are classified as thermosetting resins, and find applications in textile coatings, adhesives, and plastics, as well as paper and other areas.

At one time, urea-formaldehyde and melamine-formaldehyde resins were the predominant insolubilizers used in the paper coating industry. Urea-formaldehyde (UF) resins are no longer used to any significant extent in paper coatings. Melamine-formaldehyde (MF) resins have been replaced by other types of insolubilizers except for a few specialty applications. However, their usage is still significant.

The major disadvantage associated with these UF and MF resins is the formaldehyde odor released with their use. Formaldehyde has long been recognized as an irritant, but has recently been identified as a possible health hazard *(11–13)*. This potential hazard has resulted in government regulations directed at controlling worker exposure to formaldehyde in the workplace *(21)*. These resins contain up to one percent free formaldehyde and also evolve formaldehyde during the drying and curing operations. Other disadvantages are that an acid catalyst is required for thorough curing and an aging period is required for maximum wet-rub resistance.

Chemistry

The chemistry and technology of the various amino resins have been well covered by Wohnsiedler *(45)*, Widmer *(44)*, Petropoulos *et al.* *(46)*, and Meyer *(43)*. In recent years, progress has been made in determining the complex chemical structures present in these resinous products. Tomita *(47)* and deWit *(48)* have reported studies on the structure of UF resins using ^{13}C nuclear magnetic resonance (NMR) spectroscopy very effectively. Ebdon *(49)* has characterized both UF and MF resins using ^{15}N NMR spectroscopy.

The reactions between urea and formaldehyde are quite similar to those between melamine and formaldehyde. The higher functionality of the melamine and its properties as a weak base cause some differences in the results of these reactions. The basic reactions that occur during the manufacture of the UF and MF resins used as insolubilizers are shown in **Scheme 8.6**:

$$1)\quad R\text{-}NH_2 + H\text{-}CHO \rightarrow R\text{-}NH\text{-}CH_2\text{-}OH$$

$$2)\quad R\text{-}NH\text{-}CH_2OH + H_2\text{-}N\text{-}R \rightarrow R\text{-}NH\text{-}CH_2\text{-}HN\text{-}R + H_2O$$

$$3)\quad R\text{-}NH\text{-}CH_2\text{-}OH + HO\text{-}R^1 \rightarrow R\text{-}NH\text{-}CH_2\text{-}O\text{-}R^1 + H_2O$$

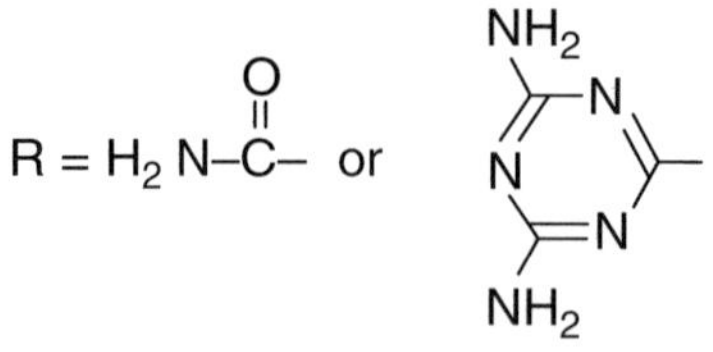

$R^1 = -CH_3, -C_2H_5, -C_3H_7, -C_4H_9$, etc.

Scheme 8.6. Reactions occurring during the formation of aminoplastresins

In reaction (1), the formaldehyde adds to the amino compound (urea or melamine) to give a methylol intermediate. In reaction (2), the methylol intermediate condenses with a free amino group forming a methylene bridge. This is the reaction that links the amino molecules together to form oligomeric or network structure. Minimizing the bridge-formation reaction in the manufacture of an insolubilizer is desirable since it results in fewer reactive methylol sites to interact with the binder (i.e., starch, protein, PVOH). Reaction (3) involves condensing an alcohol onto the methylol group. One way of accomplishing this condensation is by means of an acid catalyst using an excess of the alcohol. The excess alcohol is required to suppress the competing self-condensation reaction. After neutralization, the excess alcohol may be removed by distillation.

Commercial Use

Various types and forms of UF and MF resins have been used as paper coating insolubilizers since around 1940. Summaries of the usage of these products and their performance as insolubilizers have been given by Ritson *(14)* and may also be found in several of the TAPPI monographs *(8, 9, 50)*. Additional information may be obtained from a resin supplier *(51)*.

The early products used in this application consisted of unalkylated UF and MF resins which had been initially developed for the adhesives area. The use of UF resins waned as methylated MF resins became more popular. MF resins cured more readily and gave better wet-rub resistance than UF resins. Often a latent acid catalyst such as ammonium sulfate was employed to facilitate the cure. Neither type of resin cured very well in highly alkaline paper coatings. The MF resin — as an insolubilizer — seems to have reached the zenith of its technology with methylated MF products in the 1970s and early 1980s. Often these resins were formulated with small amounts of urea which acted as a formaldehyde scavenger.

Methylated MF resins work well at low pH, at neutral, and at moderately alkaline conditions. These resins have been successful in most major coating formulations including those bound with starch, protein, polyvinyl alcohol, and even latex binders. Often a natural binder is used in conjunction with a synthetic binder. Methylated MF resins are compatible with most coating additives and generally do not cause runnability problems. They do not cure rapidly and may require an aging period. Formaldehyde odor is one of the most serious disadvantages of these products. **Table 8.4** *(14)* shows wet-rub data under several curing conditions for two coating formulations. This data compares a UF resin, an unalkylated MF resin, and an alkylated MF resin.

Table 8.4. Comparison of urea and melamine resins on wet rub resistance with various drying conditions

Drying conditions	Wet Rub Value[a] 220°F/ 1 min	300°F/ 1 min	300°F/ 4 min
Starch-Clay Coatings[b]			
Urea-Formaldehyde	4	2.5	2
Dimethylol Melamine	4	1.5	1
Methylated Trimethylol Melamine	4	1	0.5
Casein-Clay Coatings[c]			
Urea-Formaldehyde	4.5	3	2
Dimethylol Melamine	4.5	2	1
Methylated Trimethylol Melamine	4.5	2	1

[a]*Wet Rub Resistance (Finger rub test):*
0 = No coating removed
5 = All coating removed

[b]*Coating colors: Starch — 15% on clay*
10% resin solids on starch
pH about 6–7

[c]*Coating colors: Casein — 12% on clay*
5% resin solids on casein
pH about 8–9

Depending on the performance requirements, operating conditions, binder, and the particular resin used, the amount of MF resin required may vary. For starch binders, 6–12% dry resin on dry starch is recommended *(14)*. For PVOH, 10–30% MF is recommended *(34)* (dry on dry basis). For protein, the recommended level is 5–10%.

Amino Resin Curing Mechanism

The UF or MF resins used as paper coating insolubilizers have varying proportions of the following reactive groups:

$>N-H$ (imino)

$>N-CH_2-OH$ (methylol)

$>N-CH_2-O-CH_3$ (methoxymethyl)

The methoxymethyl groups are present in methylated resins, but not to any significant degree in unmethylated resins. In general, the reactions of these functional groups in the UF resins tend to parallel those of these functional groups in the MF resins. However, UF resins tend to require more vigorous curing conditions. The UF resins might be expected to require higher temperature, more catalyst, or lower pH. The discussion on reactivity will apply to both resin types within this general framework unless noted otherwise.

The common reactive groups found in paper coating binders include the hydroxyl group of starch and polyvinyl alcohol, the amino and carboxyl groups of proteins, and the carboxyl group present in many latices. The principal reactions involved in the insolubilization process are those of the methylol or methoxymethyl groups of the resin with the reactive groups of the binder. These are illustrated in **Scheme 8.7**:

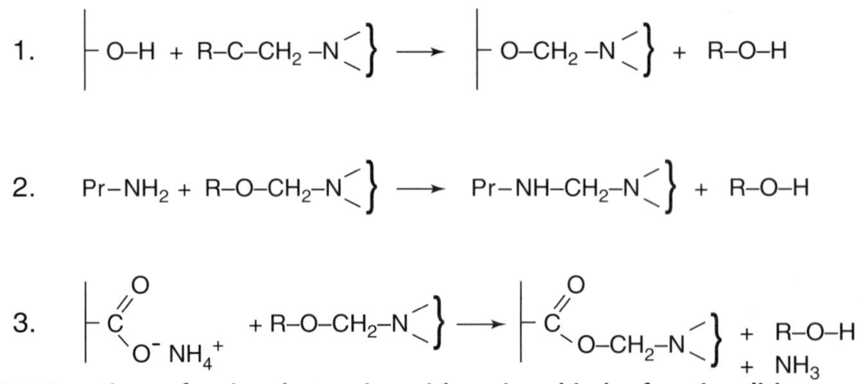

Scheme 8.7. Reactions of aminoplast resins with various binder functionalities

In reaction (1) the hydroxyl group of starch (or polyvinyl alcohol) condenses with the reactive group of the resin. If R is defined as H, as for an unmethylated resin, the condensation would release water as the by-product. If R is defined as CH_3, as for a methylated resin, the condensation would release methanol. In reaction (2), a similar condensation process occurs between the protein and the resin. In reaction (3), the carboxyl group reacts to form an ester functionality. Fully methylated MF resins are reported to react with carboxyl *(53)*, as do partially or unmethylated MF resins *(50)*.

The curing of these thermosetting resins in the presence of various polymeric binders is very complex. Little information has been published concerning these reactions under paper coating conditions. However, Petropoulos *et al. (46)* and Blank and Hensley *(53)* have studied the reactions of amino resins in aqueous industrial coatings. Their results should be somewhat applicable to insolubilizers in paper coatings. It should be recognized that there are distinct differences in curing conditions, catalysts, types, pigments, and other additives.

These authors *(53)* state that methylated urea resins are prone to self-condensation and have no reactivity toward carboxyl functionality. Partially methylolated and methylated polymeric MF resins have predominately methylol but some methoxymethyl functionality, plus reactive amino hydrogen sites. Both UF and MF resins can self-condense to evolve water, methanol, or formaldehyde. A summary of the reaction of the various groups present in these resins follows:

Reactions of the Imino Group ($\rangle$NH)

1. A site for self-condensation
2. Combines readily with methylol group
3. Not reactive with hydroxyl, carboxyl, amido, or imino groups for crosslinking

Reactions of the Methylol Group ($\rangle$N—CH_2 -OH)

1. Self-condensation under either acidic or basic conditions forming imino or ether functionality and evolving water and/or formaldehyde
2. Crosslinks under mild acidic catalysis with hydroxyl, carboxyl, amino, or amido groups.

Reactions of the Methoxymethyl Group ($>$N-CH_2 -O-CH_3)

1. Crosslinks under acidic catalysis with hydroxyl and amido groups. Methylated MF resins react with carboxyl groups; methylated UF resins do not.

2. Under basic conditions (pH 7–10), there is little if any reaction.

Based on the above information, it is probable that the methylated MF resins can crosslink with all of the common paper coating binders, including carboxylated latices. The rate and degree of reaction will depend on type of resin, pH, fugitive or fixed alkali, catalysts, drying, and curing conditions.

Glyoxal-Based or Modified Glyoxal Resins

Historical Perspective

In the 1970s, glyoxal-based resins were introduced into the paper industry, but were slow to gain acceptance. In general, these resins offered several advantages over UF and MF: no catalyst was required, no formaldehyde odor was present, and paper was cured as soon as it came off drying cans, which eliminated aging requirements.

In 1970, Ritson and co-workers patented a new type of starch insolubilizer *(54)*. This product was a low molecular weight water-soluble co-polymer of acrylamide and a small amount of acrylic acid, reacted with glyoxal. The mole ratio of glyoxal to acrylamide was 1:2. This product combined the advantages of no formaldehyde with the rapid cure of glyoxal. The examples cited in the patent used the product at a level of 10% (dry weight) based on starch, and obtained satisfactory wet-rub results. One major disadvantage of this product was its low solids formulation. The low solids formulation (15%) was necessary because of its high viscosity and tendency to gel. This product did not gain wide acceptance.

This product apparently worked by forming a moderately crosslinked polyacrylamide network which secondarily reacted the glyoxal and starch. Acrylamide/acrylic acid copolymers have been used as dry strength agents *(55)*, and this may have helped the performance of the product.

In 1973, Kelly and Lowery filed a patent on a urea-glyoxal resin that could be used at a level of 3–7% *(56)*. This product consisted of 1 mole of urea per 2 moles of glyoxal reacted together to form a mixture of low molecular weight polymers. The product was prepared at 50% solids and had excellent stability. The patent examples show that this product gives excellent wet-rub resistance, but does not adversely affect coating color viscosity as does an equivalent amount of glyoxal. The product does not appear to cure quite as fast as glyoxal.

As late as 1976, a TAPPI panel discussion on water resistance in paper coatings made several references to UF, MF, and glyoxal, but not to these glyoxal-based or modified glyoxal products. The lack of reference to these types of products would seem to indicate that they were not widely used at that time *(57, 58)*.

In the late 1970s and early 1980s, several controversial papers were published linking excessive exposure to formaldehyde vapors with health risks *(11–13)*. This information provided the impetus for several industries using formaldehyde-based resins to seek ways to reduce or eliminate exposure to this chemical. Among the affected industries was paper. In 1979, North filed a patent on a non-formaldehyde permanent press textile resin *(59)*. This resin was tried in a paper application and found to work exceptionally well. Dodd and Lane patented a paper application in 1981 *(60)*. This product was found to give wet-rub results comparable to MF resins at approximately one-half the level. Like other glyoxal-based resins, it cured on the drying cans without a catalyst, but did not adversely affect coating color viscosity. The product commercialized was a methylated ethylene urea-glyoxal condensation product, as shown in **Fig. 8.3**.

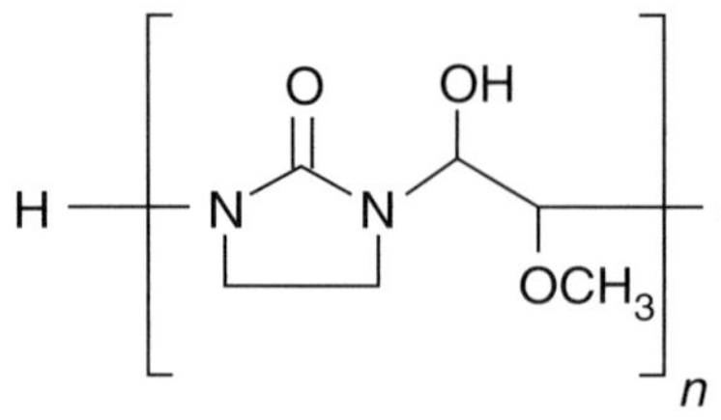

Fig. 8.3. Methylated ethylene urea-glyoxal condensation product used as a starch insolubilizer

This product was soon overshadowed by an improved version *(61, 62)* which gained wide acceptance. This new product was similar to its predecessor, but was based on a 6-membered cyclic urea. This improved product also gave excellent wet-rub values without aging, and did not adversely affect coating color rheology.

Other glyoxal-based products were used to some extent. A urea-glyoxal resin became commercially significant, as did a product that appeared to be a condensation product of glyoxal and glycols. Hui developed a polyacrylamide cyclic urea-based glyoxal resin, but was limited by low solids *(66)*. Lipowski filed a patent in 1983 for a product similar to the urea-glyoxal resin developed by Kelly and Lowery, but having urea and glyoxal in the ratio of 1 to 3 *(63)*. The examples in this patent used both all starch and starch-latex binders. Excellent wet-rub results and no adverse effects on coating color viscosity were observed, according to the patent. This product is believed to have been used commercially. Japanese scientist Kauakami and co-workers developed a polyamide-polyurea resin containing glyoxal *(69)*. Although widely used in Japan, it has not become a commercial product in the United States. Glyoxal-based resins today constitute the majority of the starch insolubilizers in use. They are commonly applied for use with natural binders in conjunction with latex.

While much work has been done developing these resins and optimizing their use in various coating compositions, little has been published on the mechanisms by which they are thought to work.

While structurally similar to North's textile resin, the starch insolubilizer in Fig. 8.3 did not need an acid catalyst (normally used in textiles) for paper applications. **Figure 8.4** shows a plot of wet-rub resistance vs. coating color pH for an alkylated ethylene urea-glyoxal resin. Optimum results are obtained at neutral or slightly alkaline conditions. This observation provides evidence that a crosslinking mechanism different from that occurring in the textiles is involved. It has also been found *(64)* that resins of this type may be treated with an aliquot of caustic and analyzed for glyoxal content the same as a glyoxal sample in a standard assay procedure *(22, 27, 28)*. This observation indicates the resin is labile under alkaline conditions and "unzips" to liberate glyoxal. Lipowski noted a similar occurrence when attempting to titrate free aldehyde groups in his polymer *(63)*. This data suggests that these products function by blocking or delaying the reactivity of the glyoxal until the heat and water removal of the drying operation allow its release. The type and reactivity of the blocking groups affect the facility with which the glyoxal is liberated.

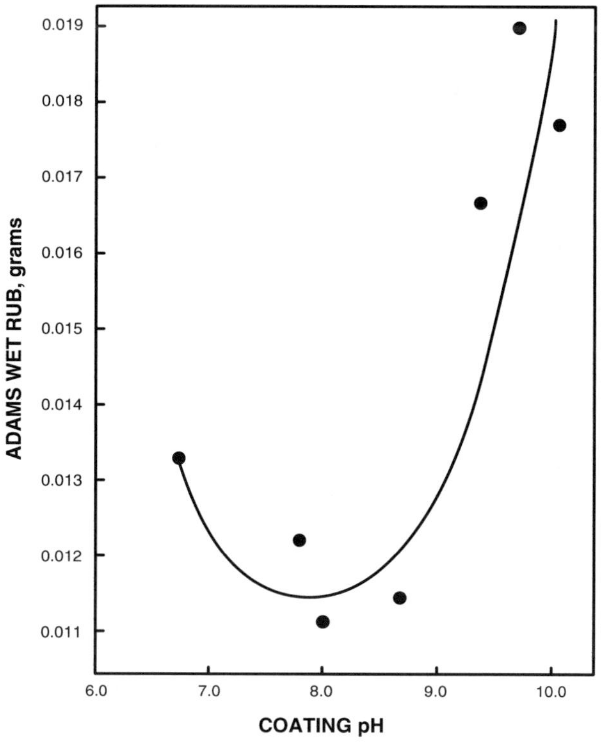

Fig. 8.4. Coating pH/wet-rub resistance (100% pigment/18% starch)

To test this blocked glyoxal hypothesis, Floyd examined non-resinous, monomeric glyoxal equivalents in insolubilizers *(65, 67, 70)*. Dihydroxydioxanes are well-characterized compounds and are known to break down in basic environments to release glyoxal *(68)*. Several of these compounds were tested and found to function reasonably well as starch insolubilizers. In an all-starch coating, these compounds exhibit a viscosity increase greater than a cyclic urea-glyoxal insolubilizer, but significantly less than an equivalent amount of glyoxal. Dihydroxydioxanes are not used commercially in paper coatings at this time, but have been mentioned as crosslinking agents in photographic gelatins *(71)*.

Methods of Use

There are several different glyoxal-based resins in use, and general statements can be made that apply to all. Much like glyoxal, these resins cure rapidly as the coating is dried on the machine. Catalysts are not needed. Insolubilization is achieved by lowering the moisture of the coating below a critical level, and not by curing at a higher temperature and retaining moisture. Aging does not significantly improve wet-rub resistance with these resins. **Figure 8.5** shows wet-rub results for an MF at 8% (dry on dry) vs. an alkylated ethylene urea glyoxal resin at 4%, dry on dry.

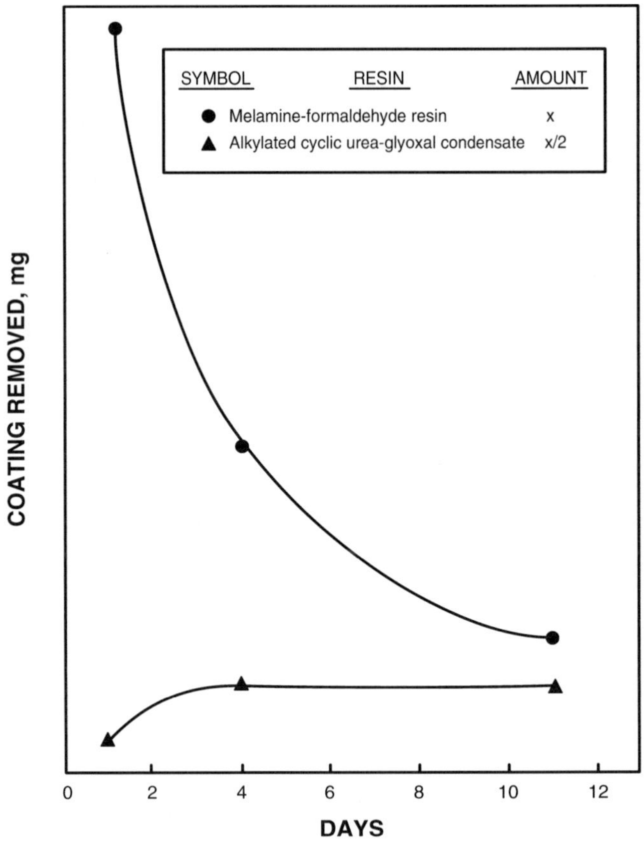

Fig. 8.5. Twenty second wet-rub test for an alkylated cyclic urea-glyoxal condensate

Glyoxal-based resins give best performance in a slightly alkaline coating. **Figure 8.4** shows wet-rub data for an alkylated ethylene urea-glyoxal resin as a function of coating pH. The wet-rub resistance is optimum at about pH 7.5–8.5. Above pH 9 or below pH 7, performance drops off rapidly. Above pH 9, the alkaline environment may destroy the glyoxal released from the resin. Below pH 7, the resin may not break down to release glyoxal.

As a first approximation for the amount of resin needed, it is generally recommended to try one-half the amount of MF resin that is being used. Thus, if an MF resin is being used at a level of 8% dry on dry starch, a modified glyoxal resin could be initially tried at 4%. The level then should be varied according to the results obtained. These resins are generally used in coatings that contain starch and various types of latex. The type of starch is also an important variable, as it is with other types of insolubilizers.

Coating formulations are prepared in the usual manner as would be done for an MF resin. The same commercial grades of ammonium persulfate converted, oxidized, or substituted starches are used with clay, calcium carbonate, or titanium dioxide pigments. Any of the common latices may be used with these resins, such as styrene-butadiene, acrylic, or vinyl acetate emulsions. The insolubilizing resin is usually added last. Viscosity increases are usually very slight in all starch binders and are lessened by addition of latex.

Several crosslinkers were evaluated in a typical recycled board, air knife coating color formulation. The formulation consisted of 13 parts latex and 7 parts of cobinder, at 45% solids. The cobinders were two different grades of soy protein, and two different grades of ultra low viscosity (ULV) PVOH, fully hydrolyzed (FH), and partially hydrolyzed (PH). Included in the formulation at 10% dry on dry parts cobinder (protein or PVOH) were: (a) glyoxal, (b) cyclic amide condensate, and (c) melamine-formaldehyde resin. The coatings were rod applied to paperboard, dried for 90 seconds at 120°C, conditioned, and tested for wet-pick resistance by the Vandercook proof press test. The results, presented in **Table 8.5**, indicate that in this formulation, the three crosslinkers were equally effective, and that the PVOH-based formulations were at least as good as the protein-based formulations *(152)*.

Table 8.5. Wet pick resistance comparisons of pvoh and soy proteins when used as coating binders (152)

Binder	Crosslinker Type (10% d/d)	Vandercook Wet Pick Resistance[a] Mills Where Picking Stopped
ULV; FH PVOH	Glyoxal	0.35
ULV; PH PVOH	Glyoxal	0.40
Soy protein A	Glyoxal	0.55
Soy protein B	Glyoxal	0.55
Soy protein B	Glyoxal	0.55
ULV; FH PVOH	CAC [b]	0.30
ULV; PH PVOH	CAC	0.40
Soy protein A	CAC	0.45
Soy protein B	CAC	0.50
ULV; FH PVOH	M-F [c]	0.35
ULV; PH PVOH	M-F	0.40
Soy protein A	M-F	0.45
Soy protein B	M-F	0.45

[a] *Scale:*

0.0 = No pick, excellent rating

1.0 = Much pick, poor rating

[b]*Cyclic amide condensate type*

[c]*Melamine formaldehyde*

Modified glyoxal resins have been used to insolubilize ethylated starch/PVOH cobinders in pigmented size-press applications. It was demonstrated that the use of insolubilizer improved the wet-rub resistance, and also eliminated a dusting problem on the calender stack. It was subsequently noted during a printing trial that picking and piling problems were observed with the control paper containing no insolubilizer, but were eliminated on the paper which contained insolubilizer *(153)*.

As these glyoxal-based resins began replacing MF resins, a new problem arose. Formaldehyde from MF resins inhibited bacterial growth in the starch-based coating color. Without the formaldehyde, bacteria — or "bug" counts — could get out of hand. Bacteria caused the starch binder to lose its adhesive properties (due to starch hydrolysis and sugar formation) and results in poor water resistance. Bug problems have caused some mills to add biocides to their formulations. Glyoxal-based resins which contain biocides are also available, and are reported to control this problem.

Several of these glyoxal-based resins have been successfully used in a size-press application. Improved sizing performance was noted using an alkylated cyclic urea-glyoxal condensate in size-press grades utilizing starch as the primary sizing agent *(72)*. The graph in **Fig. 8.6** shows that ink holdout time, as determined by the Hercules Sizing Test, increases as the amount of resin added increases. The coating contained hydroxyethylated starch with varying amounts of resin, at 10% total solids.

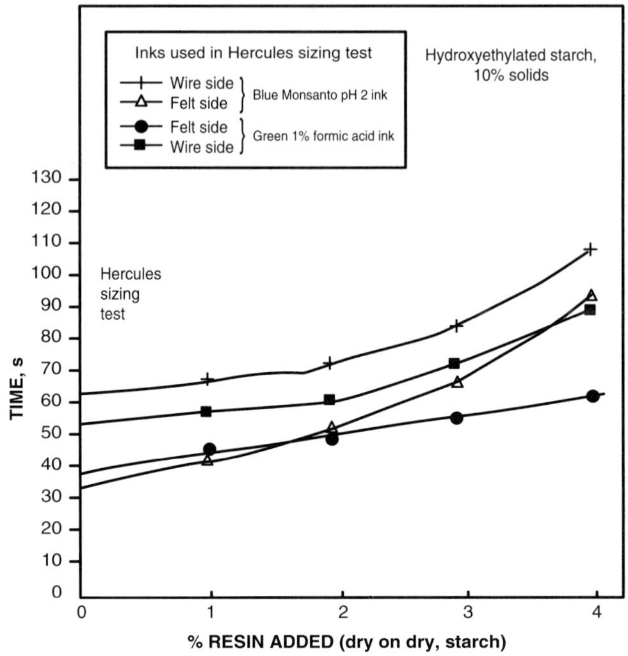

Fig. 8.6. Alkylated cyclic urea-glyoxal condensate for improved sizing in a size press coating

Simultaneous Ink and Water Application (SIWA) *(73)* test results on this paper indicated that water absorbance under dynamic conditions was reduced as the level of resin increased from zero to four percent. The reduced water absorbency indicates that dynamic water holdout is improved by incorporation of an insolubilizer in the size press.

Modified glyoxal resins have been used in protein-based coating colors. The functionality present in protein has been discussed earlier in this chapter and the reactions of glyoxal with protein may be applied here also. Reported results indicate that these resins are not as effective under typical protein formulation conditions as they are in most starch-based coating formulations *(20)*. The most likely reason for reduced efficiency is the high pH required to solubilize the protein. As mentioned earlier, blocked glyoxal resins release glyoxal under neutral or slightly alkaline conditions, and glyoxal is degraded under highly alkaline conditions. Modified glyoxal resins may also cause some color formation when used in protein-based coating which may result in a loss of brightness.

Table 8.6 *(20)* compares wet-rub performance of two modified glyoxal resins and other conventional insolubilizers used in coating colors containing hydrolyzed protein or carboxylated protein. The coating formula used was 100 parts by weight clay, 7.2 parts protein, 10.8 parts styrene-butadiene latex, and 0.3 parts insolubilizer. One of the blocked glyoxal resins was based on imidazolidinone (a 5-membered cyclic urea) while the composition of the other was unclear from the article. The wet-rub results are percent transmittance from a 10 second Adams wet-rub test.

Table 8.6. Typical results on wet rub

	Wet Rub[a], % Transmittance	
Crosslinker	Hydrolyzed PC2	Carboxylated PC4
None	60	63
Melamine	75	70
Modified melamine	65	71
Urea formaldehyde	71	64
AZC	90	88
Imidazolidinone	76	70
Glyoxal resin	69	57

[a]Adams Wet Rub, 10 seconds

Metal Salts

Historical Perspective

Over the years, many metal salts have been used to insolubilize various binder systems for paper coatings. A review of metal salts used as insolubilizer was published by Thurlow *(74)* in 1962. Zinc sulfate solutions and papermakers alum have been used to insolubilize protein-based coatings *(9, 74)*. Several metal salts, such as chromic nitrate or cuprammonium hydroxide or various dichromates, are reported to insolubilize polyvinyl alcohol. These metal salts also impart characteristic colors to the film *(34)*. Certain titanium chelates are also reported to crosslink PVOH *(34)*. Some of the metal salts that have been examined in the past are now recognized to have environmental or toxicological concerns surrounding their usage. Other literature mentions the use of chromium *(76, 77)*, antimony *(76, 78)*, zirconium *(79)*, titanium *(80)*, and tin *(81)*.

AZC

One of the most commonly used metallic salt insolubilizers is ammonium zirconium carbonate (AZC) *(84, 85)*. First used as an insolubilizer for coating binders in the early 1960s *(86)*, this product has been primarily used in protein coatings. AZC has also been used with starch, starch-latex, and all-latex binder systems. Unlike other insolubilizers, AZC is not adversely affected by high pH conditions. AZC can be used with highly alkaline pigments such as satin white or calcium carbonate. The carbonate complex in solution is unstable to heating, dilution, or pH changes. These conditions can produce a carbonated hydrous zirconia which appears as a gelatinous solid. Stability can be improved by the addition of chelating ligands such as tartrate ions *(86)*. AZC solutions are commercially available including a tartaric acid-stabilized product. These AZC solutions typically are clear, colorless, have an ammoniacal odor and generally are 20–24% active, as measured by zirconium dioxide equivalent content. The pH of these solutions is around 9.0 as received. It is reported that rapid insolubilization is obtained directly off the machine. This is because heat is not required for curing; merely the removal of water *(87)* is required. This is in contrast to the aging period often required for

MF resins. There is a potential for the release of small amounts of ammonia and its associated odor during the drying of AZC-containing coatings. Additional information may be obtained from a supplier *(88)*.

Chemistry

Ammonium zirconium carbonate is prepared as an aqueous solution of hydroxy-bridged zirconium atoms *(86)*, as shown in **Fig. 8.7**. The structure may be represented stoichiometrically as *(89)*:

$$[NH_4]_3[Zr(OH)_3(H_2O)_3(CO_3)_2]$$

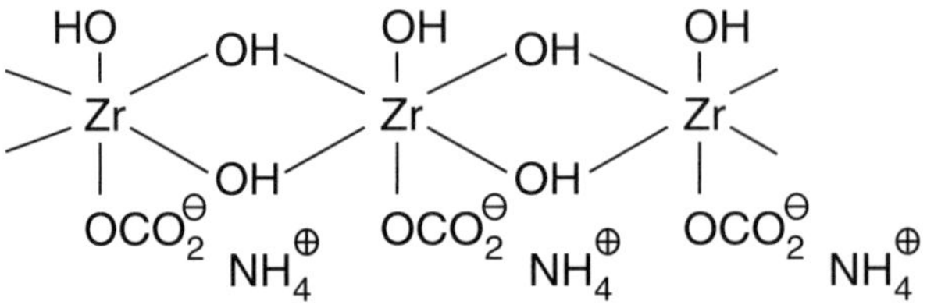

Fig. 8.7. Structure of ammonium zirconium carbonate in solution

The driving force for much of zirconium's aqueous chemistry is a strong affinity for oxygen-containing species. In particular, glycoxide, hydroxide, carbonate, alpha-hydroxycarboxylate, and carboxylate are bound most strongly, in that order *(90)*. Alcohols are weaker ligands than water. The zirconium-nitrogen bond *(91)* is considerably weaker than various Zr-O bonds.

Ammonia and organic amines do not combine with zirconium in aqueous solution through the formation of Zr-N bonds. This is because of the greater affinity of zirconium for the oxygen atoms of water than for the nitrogen atoms. This brief summary explains much of AZC's reactivity as an insolubilizer in paper coatings.

In an aqueous coating color, the zirconium's atom exists as a hydroxy-bridged, ionic oligomeric complex that is stabilized by ammonium cations and carbonate ligands *(86)*. Carbonate anion requires an alkaline environment for existence and is stable at a pH of 9–10. Under ambient conditions in a coating color, the carbonate is strongly complexed with the zirconium to the exclusion of the carboxylate functionality present.

In starch-based binder systems, there is a weak interaction between the AZC and the hydroxyl groups of the starch. This interaction is evidenced by a viscosity increase under low shear conditions. This viscosity increase can be minimized by the addition of alkali *(87, 92)* to the formulation. The AZC-starch interaction is reportedly advantageous in that it leads to increased water retention *(93)*. Increased water retention results in an improved coating structure that is more uniform in terms of wettability and ink absorbency. The ideal coating should be immobilized quickly so that little binder migration into the paper or to the surface occurs. The binder may migrate into the base sheet when the coating is applied, or may migrate to the surface when the coating is dried. Excessive binder migration can result in mottling and poor printability.

After the coating has been applied, the AZC reacts upon drying. With the heat of drying to remove the water, ammonia is lost and carbonate decomposes to lose carbon dioxide. With the loss of the carbonate ligand, the zirconium complex is reported to dimerize and react with the carboxyl groups in a crosslinking reaction which renders the binder insoluble *(89)*. The presence of carboxyl groups is required for optimal insolubilization by AZC. Thus, protein with its carboxyl groups is readily insolubilized *(95)*, oxidized starch is readily insolubilized *(87)*, and carboxylated latices are also readily insolubilized *(96)*. Polyvinyl alcohol is reported to interact with AZC via diol groups with an improvement in water resistance *(94)*.

Usage

AZC is an excellent insolubilizer for protein-based coating colors. **Table 8.5** *(20)* compares various insolubilizers for hydrolyzed protein and carboxylated protein. The coating formula used was 100 parts by weight clay, 7.2 parts soy protein, 10.8 parts styrene-butadiene latex, and 0.3 parts insolubilizer.

The graphs in **Fig. 8.8** *(87)* compare the wet-rub performance of various insolubilizers in two coating formulations. Both formulations contain 8 parts oxidized starch and 8 parts carboxylated styrene-butadiene latex as the binder. The "clay" coating contains 100 parts clay, while the "carbonate" coating contains 70 parts clay and 30 parts calcium carbonate. Both coatings are 60% solids and have a pH of 9. The graphs show that the AZC improved the water resistance of both formulations compared to the control. Thus, AZC is effective with both clay and calcium carbonate pigment systems *(87)*. AZC appears to have advantages over glyoxal in the presence of an alkaline pigment such as carbonate. AZC is normally added at a level of 5–10% (ZrO_2 basis) on the dry weight of the binder.

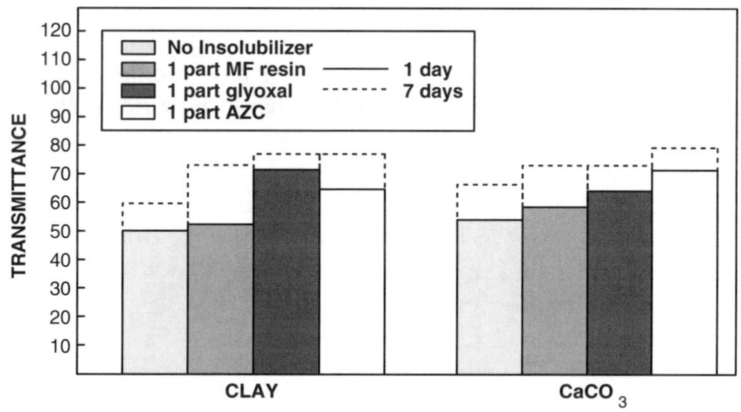

Fig. 8.8. Wet-rub resistance of 8 starch/8 latex formulations

Zirconium Chelates

A more recent development has been the introduction of ammonium zirconium chelates as protein insolubilizers *(177)*. In this type of product, the zirconium is chelated with lactic acid, citric acid, or a blend of these acids. These hydroxyacids stabilize the zirconium over a pH range of 6.5 to 10. Varying the ratio of lactic acid to citric acid affects the rate of reactivity of these compounds. These chelated products are prepared at a pH of 7 to 8, which results in less free ammonia than found in traditional AZC-type products *(178)*. These chelated products are reportedly more efficient in that they don't dimerize when cross-linking *(86, 89)*. Although they are formulated at a 7% ZrO_2 basis vs. 20% for the AZC types, this high efficiency allows the recommended usage level to be the same wet weight as for an AZC-type product *(178)*.

Zinc Salts

Zinc sulfate solutions have been used commercially to insolubilize protein. These solutions are sometimes applied subsequent to the coating application. The protein-containing coated paper may be dipped in a bath or kiss coated with a roll coater. Zinc formate and zinc acetate are mentioned in the patent literature as insolubilizers in the pH region of 5.7 to 6.2 for protein coating colors which coagulate upon heating *(145)*. A proprietary ammonium-stabilized zinc complex *(147)* is used commercially as a protein insolubilizer in high pH coating colors. Zinc salts may have an economic advantage over zirconium slats and are reported to be effective at lower levels of usage *(146)*. Zinc salts are also reported to require some diligence in their usage and formulation. Coating color instability and rheology problems may be encountered *(146)*. Prestabilization of the system with dispersants or nonionic surfactants may be required prior to introducing the zinc complex *(148)*. Urea and dicyandiamide have been used to reduce the viscosity of proteinaceous binder formulations when zinc salts are present *(145, 149, 150)*. A typical coating formulation using a zinc complex insolubilizer might be as follows (on a dry basis) *(146)*:

94 parts	clay or pigment
6 parts	soy protein
6 parts	carboxylated styrene-butadiene latex
0.48 parts	zinc complex (as zinc metal)
1 part	hexametaphosphate dispersant

This would be formulated at a pH of 9–10, with 40–44% solids for an air knife coater or 55–60% solids for a trailing blade coater. The use of zinc compounds in high pH proteinaceous coating colors is interesting in that the insoluble zinc hydroxide readily precipitates as acidic aqueous zinc salt solutions are neutralized. In the presence of strong base, zinc hydroxide dissolves, forming the zincate ion, commonly written as $ZnO^{=}_2$ *(151)*. In the presence of strong ammonium hydroxide solutions, the soluble amine complex is found: $Zn\ (NH_3)^{2+}_4$ *(151)*. When the coating is dried and cured, the zinc complex breaks down, releasing ammonia and water. The zinc then reacts with carboxylate groups in a crosslinking reaction.

Reactive Starches

An interesting development in the insolubilization of starch and other binders has been the modification of starch so that the starch itself becomes the reactive crosslinking agent. One of the earliest such products was dialdehyde starch (DAS). Other approaches have been to attach a pendant acrylamide functionality or a maleic acid ester. Other workers have improved the insolubilization by making an amphoteric starch complex.

Dialdehyde starch is prepared by the periodic acid oxidation of regular starch *(97, 98)*. Periodic acid selectively effects the oxidative cleavage of 2,3-diol group of the anhydroglucose ring in starch to yield the dialdehyde. This reaction is shown in **Scheme 8.8** *(99)*.

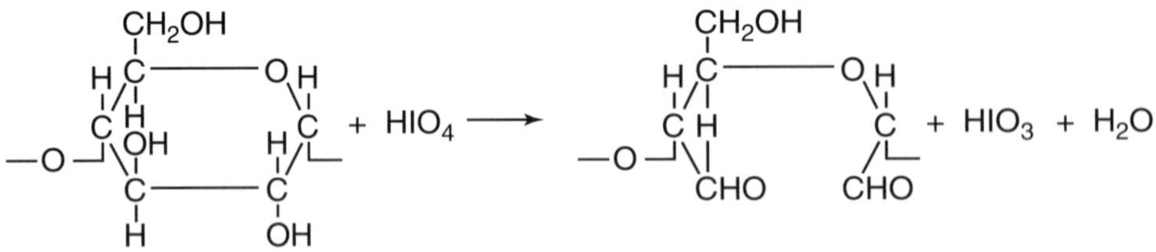

Scheme 8.8. Preparation of dialdehyde starch

An electrolytic cell process using and recycling iodine and iodic acid to create periodic acid has been developed for commercial production *(100–102)*. The oxidation is readily controlled so that products with any level of dialdehyde content may be prepared. Despite the name of this product, it contains very few free aldehyde groups. These reactive groups combine with the proximate starch hydroxyls to form intra- and intermolecular hemiacetals, or exist in a hydrated form *(103–105)*. A comprehensive coverage of dialdehyde starch has been given in various reviews *(99, 103, 104, 106, 107)*. Large-scale consumption of DAS has not materialized because of its relatively high cost *(108)*.

Weakley *et al.* reported that DAS will react with casein to give an insoluble product *(109)*. Ernst *et al. (110)* and Weakley and Mehltretter *(111)* disclosed the preparation of a copolymer of DAS and a protein that provided wet-rub resistance in pigmented paper coatings. Carr *et al. (112)* demonstrated that dialdehyde starch with a 10–60% degree of oxidation can be used as the sole adhesive for a wet-rub resistant paper coating. Zienty *(113)* prepared a wet-rub resistant starch-clay coating by applying a top coating which consisted of a dispersion of DAS combined with a dicyandiamide-formaldehyde reaction product.

Suzumura and Miyahara *(114)* disclosed paper coating composition of various binder combinations based on the use of DAS and polyvinyl alcohol. Curtis *(115)* indicated that, in a casein-latex formulation, part of the casein could be replaced by a mixture of ethylated starch and cationic DAS. Weakley *et al.* studied the use of DAS as an insolubilizer for soy protein coatings *(116)*.

Young and Verbanac developed a starch which contains pendant acrylamidomethyl ethers by condensing N-methylol acrylamide onto starch hydroxyl groups *(117, 118)*. These pendant acrylamide groups could be copolymerized with other ethylenically unsaturated monomers present. The pendant acrylamide groups could also be induced to crosslink with themselves via homopolymerization under conditions of ultraviolet light, electron beam radiation, thermal initiator, or by an alkali-catalyzed addition reaction at elevated temperatures *(119)*. Films cast from this composition were water-insoluble after curing. The water resistance of these films was greatly enhanced by UV curing in the presence of thermally cured crosslinking resins. Films incorporating co-monomers such as hydroxyethyl acrylate were found to have good flexibility, indicating that the monomers create internal plasticization *(120)*. After crosslinking, this starch is reported to show significantly increased resistance to the penetration of both oil and water into the coated surface. In addition to increased barrier properties, strength and flexibility properties are enhanced when this product is used in paper coating or sizing applications *(121)*. A similar product was developed by Tessler *(122)*.

Zimmerman developed an amphoteric starch having a net anionic charge by reacting maleic anhydride with a cationic starch having quaternary ammonium groups *(123)*. This starch was designed for use with melamine-formaldehyde insolubilizers in a pH range of 6 to 9.5. The pH of such a coating was adjusted with ammonia, which flashed off during the drying. The resulting pH drop could regenerate the acidic functionality of maleic groups which could catalyze the reaction of the MF resins. This starch is claimed to be capable of self-crosslinking at a high pH to produce water-resistant insolubilized coatings.

Hinterwaldner *(124)* examined acrylic modified polymerizable starches. These products were used in crosslinkable coatings.

Paschall *et al. (125)* formed a starch propionamide which was reacted with formaldehyde. Thayer *(126)* used starch propionamide and reacted it with polyaldehydes to get wet-rub resistance under alkaline conditions.

Cescate *(127)* prepared an amphoteric starch derivative by first attaching cationic groups to the starch and then oxidizing the cationic starch to provide anionic groups.

The resulting amphoteric starch was used with glyoxal as the insolubilizer to provide wet-rub resistance. Taylor *(128)* of Staley used an anionic starch in conjunction with polyalkyleneimine to enhance its performance as a pigment binder.

Aziridinyl starch derivatives have been prepared under carefully controlled conditions. These starches can be reacted further and are useful as latent crosslinking agents *(129, 130, 137)*.

Other Types of Insolubilizers

Various other types of insolubilizers have been patented over the years, and some may have attained commercial usage on a limited basis. Some are reportedly used abroad to some degree, but are not currently in use in the United States.

Ritson and Jen *(131)* described a cationic thermosetting resin. This resin was prepared by reacting a salt of a polyamine with a water-soluble metal cyanate followed by reaction with formaldehyde. Tanaka and Nonaka *(132)* used an epichlorohydrin-polyamide reaction product in starch coatings containing alkaline pigments. Strother and Coker *(133)* have disclosed starch-clay coatings with improved water resistance when a copolymer of a polyalkyleneimine and epichlorohydrin was added. Sumitomo Chemical Corp. has developed a series of polyamide-polyamino resins which are reacted with formaldehyde, glyoxal, or epichlorohydrin *(134, 135)*. Nakanishi *et al.* of Asahi-Dow Ltd. *(136)* disclosed a binder system containing a carboxylated styrene-butadiene latex which is crosslinked by a polyfunctional aziridine compound.

Kress *(138, 139)* claimed the use of polymeric acetals for insolubilizing both starch and protein binders. Perkins *(140)* used diphenolic acid as an insolubilizer. McKnight and Brown *(141)* used monomethylol dimethylhydantoin and similar products which functioned as formaldehyde donors for insolubilization. Other patents have been mentioned in a previous monograph *(2)*.

Printing Demands Affecting Insolubilizer Needs

Offset is still the number one printing process as of 1990, representing a US$ 65 billion market of finished goods in the United States, with flexography at US$ 40–45 billion, gravure at US$ 14 billion, and screen and letterpress at US$ 3 billion *(154)*. Technical innovations and environmental regulations continue to drive the changes in the printing industry. Most of these changes are the result of increased press speeds, higher quality print requirements, and the need to reduce solvent emissions.

Water-based gravure and flexo systems have been successful and are gaining in popularity. With more water on press, greater demands are being placed on the paper performance. Since water-based systems are now in their incipiency stage in publication gravure markets, it is unclear exactly what type of long-term coating insolubilization parameters must be met. The challenges that are already being recognized in this water-based gravure market are picking, water resistance, fiber puffing, and drying limitations. The newsprint market is starting to move away from the offset process and going to water-based flexo systems. Fiber lift and plate plugging are the two biggest concerns with the flexo units printing water-based news in the United States, but improvements are being made *(155)*.

In addition to meeting the challenges of changing printing industry requirements, the same grade of coated paper may have to provide acceptable print performance in a number of different printing processes. Many times the papermaker will not know in advance which printing process will be selected, so it is common practice to use insolubilizers to make the paper or board suitable for offset printability.

The water resistance developed in a paper coating with the use of insolubilizers is particularly critical in the web offset process where aqueous-based dampening solutions are employed. If the binders in the coating formulation, especially natural binders, are not properly insolubilized, piling on press will result.

Printability Results

Increasing press speeds and consequent changes in ink chemistry and fountain solutions necessitate reassessment of the nature of coated paper and board surfaces. Traditionally, coatings have been designed to achieve high levels of wet-rub resistance. Such is no longer the case. High-speed printing processes require rapid acceptance of aqueous and oily fluids for high quality printing.

A suitable balance of absorptivity and wet-rub resistance must be achieved to maintain printability. On press, ink is an emulsion of ink and water, the latter being primarily on the surface of the ink droplet. Water, being incompressible, must escape when pressure is applied. If the water cannot move vertically, it must move horizontally and will therefore interfere with the nearly simultaneous application of the ink.

Inks are also a delicate balance of pigments in a multi-solvent system. Part of the setting/drying mechanism of ink depends on upsetting this balance. Absorption of key "balancing" solvents is critical if the ink dot is to be set properly on the coated surface. Should this absorption not occur, the dot will remain fluid and be susceptible to reverse trapping at the next station. Insolubilizers can often contribute to achieving the proper ink and water receptivity needed for better halftone dot definition and sharp, clear printed images *(156)*.

Testing Methods

Surface properties of coated paper and board are critical parameters in determining the print quality potential of any given substrate. Requirements for good printing varies, in some respects, for the offset process versus letterpress and gravure grades. In offset printing, higher tack inks are used, making important surface strength and resistance to dry picking and, also, the need for adequate internal bond of the base sheet. The use of water solutions on the printing plate and blanket adds extra concerns involving adverse effects on the offset press.

It is important to realize that in the offset process every part of the paper surface is stressed by either an inked surface or a wetted blanket surface. With these types of conditions, one of the most common difficulties encountered in the offset process is the weakening or softening of the coating binder. This can result in wet picking or contamination of the blanket and printing plate (piling).

Both "wet" and "dry" testing procedures are discussed in this chapter. Dry conditions are important to the discussions of insolubilizers as these products improve dry surface properties, as well as the wet property conditions encountered on press. Some of the same test equipment may be employed in several of the various procedures outlined, but may use alternative test conditions to simulate different press conditions more realistically.

The best procedure to evaluate paper designed for offset printability is on an actual offset press using rigorous test conditions. In recent years, the role of an insolubilizer on printability has focused on its effect on water and oil receptivity of the surface, on wet rub improvements, and on dry surface improvements. Actual press signatures allow for the comparison of halftone dot definition and image clarity differences with the choice of insolubilizer. The photomicrographs presented on the next page illustrate the importance of an actual press run to evaluate the paper from an insolubilizer trial. The first set of photomicrographs in **Fig. 8.9** is from a lightweight-coated trial comparing a starch/latex coating formulation with and without insolubilizer. The formulation with insolubilizer ran with cleaner blankets on press and exhibited better halftone dot definition to provide a cleaner, sharper printed image. The second set of photomicrographs in **Fig. 8.10** is from an insolubilizer trial in an all-synthetic binder formulation. Both formulations exhibited excellent wet-rub and press runnability, but the formulation containing insolubilizer produced better halftone dot definition. The formulations used are shown in **Table 8.7** and **Table 8.8** *(156)*.

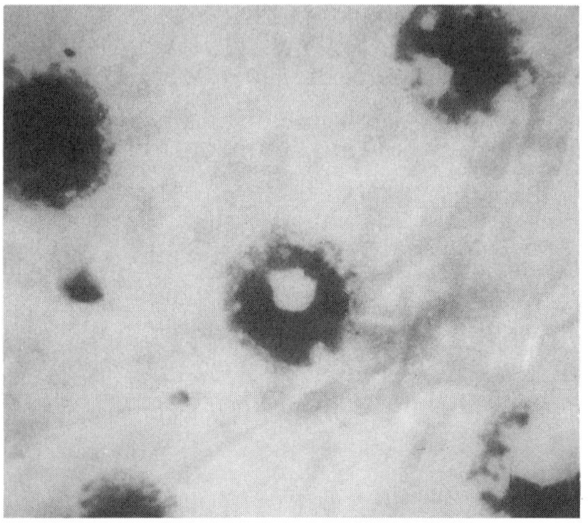

200x No Insolubilizer Cobinder System

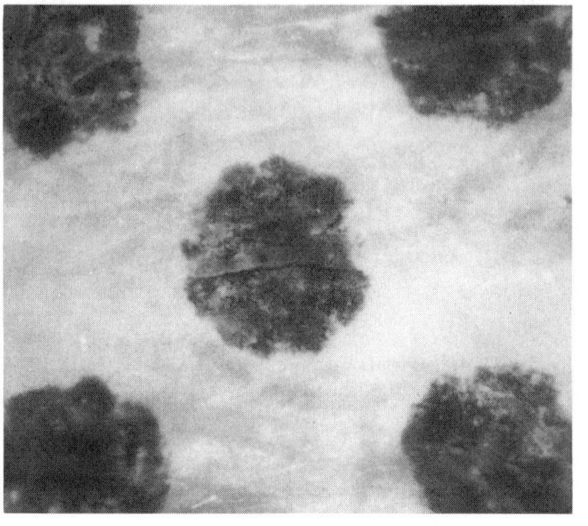

200x Cyclic Urea-glyoxal Insolubilizer Cobinder System

Fig. 8.9. Photomicrograph 1 ***Photomicrograph 2***

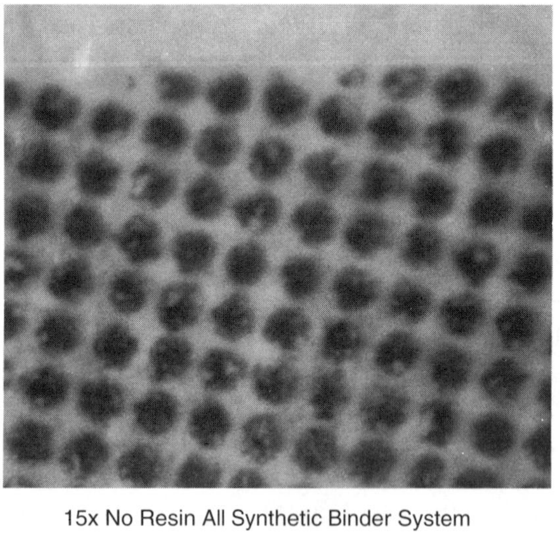

15x No Resin All Synthetic Binder System

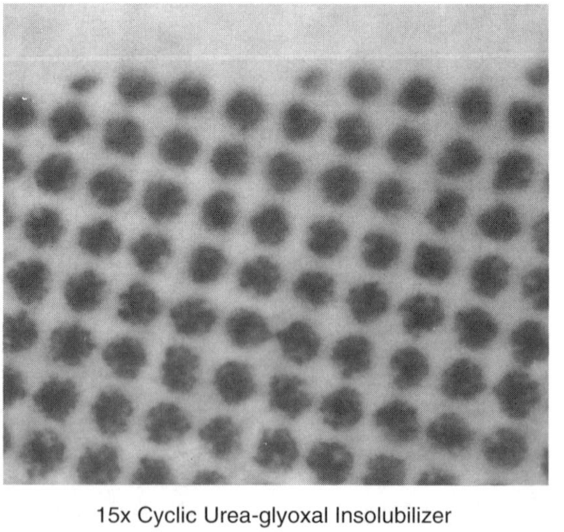

15x Cyclic Urea-glyoxal Insolubilizer All Synthetic Binder System

Fig. 8.10. Photomicrograph 3 ***Photomicrograph 4***

Table 8.7. Formulation based on 100 parts pigment

		Photomicrographs	
		1	2
No. 1 Clay	50 parts		
Delaminated Clay	50 parts		
	100 parts		
Dispersant	0.15		
Pearl Starch	9.0		
SBR Latex	8.0		
Cyclic Urea-Glyoxal Insolubilizer		0	0.36
Calcium stearate Dispersion	0.75		

Solids = 61%
pH = 8.0

Table 8.8. Formulation based on 100 parts pigment

		Photomicrographs	
		3	4
No. 1 Clay	64 parts		
High Gloss Clay	28 parts		
Plastic Pigment	8 parts		
	100 parts		
Dispersant	0.15		
SBR Latex	14.0		
Cyclic Urea-Glyoxal Insolubilizer		0	0.50
Calcium stearate Dispersion	1.0		
Sodium Polyacrylate Thickener	0.2		

Solids = 62%
pH = 8.0

Unfortunately it is not always practical to make everyday quality decisions with an actual press. Commercial press time is expensive, and it is difficult to maintain an adequate, readily available supply of paper for testing. Laboratory procedures have been developed which mimic or exaggerate various aspects of the printing process.

In the following sections, a number of evaluation methods are reviewed that can be used in the laboratory to test the effectiveness of an insolubilizer and to predict press performance. Many of these procedures can be found in the TAPPI Technical Information Sheets *(157)*, while others have been more recently developed by research and supplier organizations but have not been standardized throughout the paper industry.

Tests on Dry Paper

Dry Pick

The most widely used laboratory procedure is the IGT Printability Tester described in TAPPI procedures *(158)*. The IGT unit uses a printing disc that can be inked with tack-graded inks or viscosity-rated butane oils and printed onto the paper surface using standard conditions. By varying the speed, a point is reached where pick failure occurs providing an indication of the surface strength.

A general dry-pick resistance method still used in the industry is the (Dennison) Wax Pick Test *(159)*. This test uses a graded series of sealing waxes of differing compositions and melting points. Traditionally the Dennison waxes have been used with papers containing starch or protein binders. With the use of higher percentages of thermoplastic latex binders in paper coating formulations, the use of molten waxes for testing does not give satisfactory results. There are also other special devices designed for pick testing such as the Graphic Arts Technical Foundation Tester *(160)* and the Interfiber Bond Tester *(161)*. Many paper producers use a proof press to evaluate dry-pick performance.

Regardless of which test method is employed, it must be considered that day-to-day variations in laboratory technique and conditions are encountered. Therefore, it is best to use a relative ranking system to evaluate dry-pick parameters with the control being a standard substrate that is known to have printed satisfactorily on press.

Dry Rub

There are several methods available for the evaluation of abrasion or rub resistance of dry surfaces. The most popular method is the Sutherland Tester. This tester is used to determine the dry rub or wet rub resistance of ink films on paper. *(162)*. Another method includes a simple device described by Berilla in which the dusting tendency of a coated paper is evaluated on an empirical basis *(163)*. Also, the textile crockmeter can provide a relative ranking of dry-rub performance *(164)*.

Adsorptivity Testing

Often it is necessary to determine the effects of an additive on the ink and water absorptivity of the coated surface. The surface absorptivity is a delicate balance if good, clean printability is to be achieved. Using the IGT Printability Tester, dynamic water and Magie Oil testing can be performed to predict the influences of a particular additive on these properties *(165)*. An inclined plane test can also be employed for absorptivity determination as well as more detailed contact angle measurements using fluids of particular interest *(166)*.

Printability Evaluations

Several test procedures are available on a laboratory scale that are more sophisticated in their method to predict offset printability. One such evaluation uses the AIC2-5 model of the IGT Printability Tester to print an ink and water film simultaneously to simulate offset press conditions and speeds. Ink film thickness and the amount of water or fountain solution can be controlled to give consistent results. This test method is referred to as "SIWA" which is an acronym representing Simultaneous Ink and Water Absorption. The printed SIWA test strip can be evaluated for print uniformity in addition to absorptivity and receptivity properties *(167)*. A Proof-Bau Tester can also be used to print two colors on a lab scale while maintaining simulated press speeds. More details on this procedure can be obtained from the equipment supplier *(168)*.

Test on Wetted Paper

Many of the test methods discussed in this section are extensions of the dry surface procedures already described.

Wet-Pick

Several wet-pick procedures using the IGT Printability Tester have been outlined by supplier literature and by individuals who have made modifications to the supplier's procedure *(169, 170)*. This particular wet-pick test seems to give less consistent results versus the dry-pick test using the IGT unit due to the increased number of variables introduced.

There is also a TAPPI Useful Method (Method 513) *(171)* that describes a special attachment for the Vandercook proof press to simulate dampening on an offset press.

Wet-Rub

There are a number of test procedures used throughout the paper industry to evaluate wet-rub performance. Most of these test procedures are more simple to run than the wet pick tests and are more empirical in nature. It is therefore best in any of the methods described in this section to use a control and to evaluate the data based on relative rankings versus the control. This test parameter is most critical in evaluating the performance of an insolubilizer and its primary function of eliminating or minimizing press piling.

The Adams wet-rub tester has gained popularity in the last ten years due to its reproducible results between individual testers. It is an improvement over the more difficult to standardize wet finger test methods, such as the TAPPI Useful Method 468, Wet Finger Test *(172)*. The Adams test involves a moistened rubber roll running against a sample mounted to a backing roll for a specified time and pressure. The coating removal is measured gravimetrically to determine milligrams of coating removed or by turbidity measurements of the water suspension in which the rubber roll is immersed *(173, 174)*.

Another method used in the industry is the Taber Abrasion Test *(175)* which employs a sample on a turntable, immersed in water with the surface being rubbed by a stationary paint brush. The amount of coating removed in a set time frame is expressed by the turbidity of the water suspension.

The textile crockmeter can also be used under wet conditions to empirically determine the wet rub of a particular coated surface *(164)*. A Tri-Pick Tester has also been mentioned for wet-rub evaluations *(176)*.

Testing Needs

The previous sections outlined a number of test procedures that can be used to evaluate the performance of insolubilizers and to determine their potential effect on paper performance under press conditions. Some of these methods are more standardized than others, but they have all been effectively used as indicators of paper quality and end-use performance. Most organizations have learned to rely on a battery of test methods to get a truer picture of paper properties and to judge the performance of any particular substrate for the end-use conditions that may be encountered. Following paper with known data from a battery of tests and selecting control papers from actual press runs will greatly aid in the interpretation of laboratory results. However, as indicated earlier, the commercial press will always be the final judge of performance.

As a wide variety of final end-use conditions may be encountered, the more testing procedures performed, the better the paper can be judged relative to how severe an adverse condition it may be able to tolerate. Since the paper coating manufacturer does not have control over the various applications to which his paper may be subjected, the use of an insolubilizer has always been an excellent safety factor to provide adequate water resistance properties for a variety of applications.

Literature Cited

1. Hall, H. R., Ed., *Paper coating additives,* Monograph No. 25, TAPPI PRESS, Atlanta, 1963, Chap. 7, pp. 99–116.
2. Landes, C. G., *et al.*, in *Paper coating additives,* 2nd edn.; Landes, C. G. and Kroll, L., Eds., TAPPI Monograph No. 25, TAPPI PRESS, Atlanta, 1978, Chap. 9, pp. 136–172.
3. Nadleman, A. H. and Baldauf, G. H., *Coating Formulations — Principles and Practices,* Lockwood Trade Journal Co., New York, 1966.
4. Loomer, J. T., "In Coated Papers" (Britt, K. W., Ed.), *Handbook of Pulp and Paper Technology,* 2nd edn., Van Nostrand Reinhold, New York, 1970, pp. 517–528.
5. *Pulp and Paper — Chemistry and Chemical Technology,* 2nd edn. (Casey J. P., Ed.), Interscience, New York, 1961, Vol. III.
6. Gunning, J. R., in "Coating" (MacDonald, R. G. and Franklin, J. N., Eds.), *Pulp and Paper Manufacture,* 2nd edn., McGraw-Hill, New York, 1969, Vol. II, Chap. 9, pp. 457–532.
8. Kearney, R. L. and Mauer, H. W., Eds., *Starch and Starch Products in Paper Coating,* TAPPI Monograph No. 17, TAPPI PRESS, Atlanta, 1990.
9. Strauss, R., Ed., *Protein Binders in Paper and Paperboard Coating,* TAPPI Monograph No. 36, TAPPI PRESS, Atlanta.
10. Sinclair, Alvin R., Ed., *Synthetic Binders in Paper coatings,* TAPPI Monograph No. 37, TAPPI PRESS, Atlanta, 1975.
11. Swenberg, J. A., *et al., Cancer Research,* 1980, 40, 3398–3402.
12. Waritz, R. S. and Clary, J. C., Project Number 79-7259, "A 26-Week Inhalation Toxicity Study of Formaldehyde in the Monkey, Rat and Hamster," Formaldehyde Institute, Scarsdale, NY, April 19, 1982.
13. "Chemical Regulation Reporter," Current Report, Bureau of National Affairs, Washington, D C, p. 1160, October 19, 1979; p. 1212, February 26, 1979.
14. Ritson, D. D., *Tappi* 46(10): 600–605(1963).
15. Walker, J. F., "Formaldehyde," in *Kirk — Othmer Encyclopedia of Chemical Technology,* 2nd edn., John Wiley & Sons, New York, 1966, Vol. 10, pp. 77–97.
16. Walker, J. F., *Formaldehyde,* ACS Monograph No. 159, 3rd edn., Reinhold, New York, 1964.
17. *Chemistry and Industry of Starch,* 2nd edn. (Kerr, R. W., Ed.), Academic, New York, 1950.
18. Radley, J. A., *Starch and Its Derivatives,* 4th edn., Chapman & Hall, London, 1968.
19. Whistler, R. L., and Paschall, E. F., *Starch Chemistry and Technology,* Vols. I and II, Academic, New York, 1965 and 1967.
20. Coco, E. C., "Soy Polymer Insolubilization," Ralston Purina Company, St. Louis, MO, 1986.
21. "Final Rule on Workplace Exposure to Formaldehyde," (29 CFR 1910-1048, FR Vol. 52, No. 233), Occupational Safety and Health Administration, Washington, DC, 1987.
22. "Glyoxal," American Hoechst Corp., Somerville, NJ, 1984.
23. Buttrick, G. W., Kelly, G. B. Jr., and Eldred, N. R., *Tappi* 48(1): 28–33(1965).
24. Buttrick, G. W., and Eldred, N. R. *Tappi* 45(11): 890–893(1962).
25. Tsunemitsu, K., *et al.*, in *Polyvinyl Alcohol — Properties and Applications* (Finch, C. A., Ed.), John Wiley & Sons, New York, 1973, pp. 277–318.
26. Colgan, G. and Plante, P., in *Polyvinyl Alcohol,* TAPPI Monograph No. 25, TAPPI PRESS, Atlanta, 1963, Chap. 8, pp. 117–136.
27. "Glyoxal — A Technical Brochure," American Cyanamid Co., Wayne, NJ, 1985.
28. "General Chemistry of Glyoxal," Union Carbide Corp., New York, NY, 1967.
29. Whipple, E. B., *J. Am. Chem. Soc.* 92, 7183(1970).
30. Kliegman, J. M., *et al., J. Org. Chem.* 37, 1276(1972).
31. Chastrette, F., *et al., Bull. Soc. Chim. Fr.* 1–2, 1133–40(1983).
32. Saloman, P., *Acta Chem. Scand.* 10(2): 311(1956).

33. Arcus, C. L. and Jackson, B. A., *Chem. and Ind.* 2022(1964).
34. "Increasing Water Resistance of ELVANOL® Polyvinyl Alcohol," DuPont Co., Wilmington, DE, 1987.
35. Mazzarella, E. D. and Hickey, L. J., *Tappi* 49(12): 526–532(1966).
36. Hiemstra, P. and Vandermeeren, J., *Tappi* 52(2): 263–269(1969).
37. Harsveldt, A., Smit, G., and Zijderveld, A. H., *Paper Technology* 9(3): 204(1968).
38. Loomer, J. T., U. S. Patent 2,632,714, 1953.
39. Noel, L. W. and Wolfe, D. L., U. S. Patent 2,860,073, 1958.
40. Vail, S. L., *J. Org, Chem.* 27, 1071(1962).
41. Schwalbe, H. C., Hydell, R. C., and O'Connor, J. J., U. S. Patent 2,700,621, 1955.
42. Ritson, D. D. and Landes, C. G., *Tappi* 38(5): 300–303(1955).
43. Meyer, B. *Urea-Formaldehyde Resins*, Addison-Wesley, Reading, MA, 1979, Chaps. 3 and 4.
44. Widmer, G., "Amino Resins" (Bikales, N. M., Mark, H. G., and Gaylord, N. G., Eds.), in *Encyclopedia of Polymer Science and Technology*, Interscience, New York, 1965, pp. 1–94.
45. Wohnsiedler, H. P., "Amino Resins and Plastics" in *Kirk-Othmer Encyclopedia of Chemical Technology,* John Wiley & Sons, New York, 1963, Vol. 2, pp. 225–258.
46. Petropoulos, J. C., *et al.*, "Chemistry and Technology of Amino Resins" in *Applied Polymer Science* (Craver, J. K. and Tess, R. W., Eds.), Organic Coatings and Plastics, Chemistry Division of American Chemical Society, Washington, DC, 1975.
47. Tomita, B. and Hatono, S., *J. Pol. Sci. Polym. Chem. Ed.* 16, 2509–2525(1978).
48. deWit, J., *et al.*, *Angew. Makromol. Chem.* 62, 7–31(1977).
49. Ebdon, J. R., *et al.*, *Polymer* 25, 821–825(1984).
50. Silvernail, L. H. and Bain, W. M., Eds., *Synthetic and Protein Adhesives for Paper Coating,* TAPPI Monograph No. 22, TAPPI PRESS, Atlanta, 1961.
51. Some suppliers of melamine-formaldehyde resins are: (a) American Cyanamid Co., Wayne, NJ; (b) Monsanto, St. Louis, MO; and (c) Sequa Chemical Corp., Chester, SC.
52. "Cymel® 303 Cross-linking Agent," American Cyanamid Co., Wayne, NJ, 1979.
53. Blank, W. J. and Hensley, W. L., *J. Paint Tech.* 46(593): 46–50(1974).
54. Cosica, A. T., Ritson, D. D., Kinnelon, C. F., and Ross, J. H., U. S. Patent 3,549,568, 1920.
55. MacDonald, R. G., Franklin, J. N., *Papermaking and Paperboard Making,* 2nd edn., McGraw-Hill, New York, 1970, Vol. III, pp. 58–67.
56. Kelly, G. B. and Lowry, R. H., U. S. Patent 3,869,296, 1975.
57. (a) Rowsam, K. P. Jr., "Use of Crosslinkers to Attain Water Resistance" in *Water Resistance in Paper Coating — A Panel Discussion* (Athey, R. D., Ed.), TAPPI PRESS, Atlanta, GA, 1979, p. 17; (b) Daly, T. W. *Insolubilization of Proteinaceous Binders, ibid.,* p. 9; (c) *Panel Question and Answer Session, ibid.,* pp. 31–35.
58. Shirt, R., *Starch Effects on Water Resistance of Paper coatings, ibid.,* p. 7.
59. North, B. F., U. S. Patent 4,248,758, 1981.
60. Dodd, J. A. and Lane, R. I., U. S. Patent 4,343,655, 1982.
61. Floyd, W. C. and North, B. F., U. S. Patent 4,455,416, 1984.
62. Floyd, W. C. and North, B. F., U. S. Patent 4,505,712, 1985.
63. Lipowski, S. A., U. S. Patent 4,471,087, 1984.
64. Floyd, W. C., Sun Chemical Corp., unpublished observation, 1984.
65. Floyd, W. C., U. S. Patent 4,537,634, 1985.
66. Hui, S. H., U. S. Patent 4,544,609, 1985.
67. Floyd, W.C., U.S. Patent 4,547,580, 1985.
68. Venuti, M. C., *Synthesis,* 1982, pp. 61-63, and references cited therein.
69. Kawakami, S., *et al.,* Japanese Patent 84129229, 1984; Chem. Abstr., 1985, Vol. 101 (22), Sect. 143, Abstr. No. 194023.

70. Floyd, W. C., U. S. Patent 4,656,296, 1987.
71. Yamamoto, N. and Ashigara-Machi, M., U. S. Patent 3,847,948, 1974.
72. Wimme, M. and Foster, H. S., Technical Service Dept., Sun Chemical Corp., unpublished results, 1983.
73. Sunrez® 700 Sales Manual, "Coated Paper Testing," Sun Chemical Corp., 1984.
74. Thurlow, J. F., "Recent Developments in the Improvement of Wet-Rub Resistance in Starch Coatings," Coating Research Bulletin, Gorham Laboratories, Gorham, ME, 1962.
75. Ritson, D. D., *et al.,* in *Paper coating additives,* TAPPI Monograph No. 25, TAPPI PRESS, Atlanta, 1963, Chap. 7, pp. 95–116.
76. Iler, R. L., U. S. Patent 2, 359,858, 1944.
77. Caeser, G. V., U. S. Patent 2,111,202, 1938.
78. Griffin, I., *et al.,* U. S. Patent 2,413,320, 1946.
79. Taylor, K. G., U. S. Patent 3,081,199, 1963.
80. Taylor, K. G., U. S. Patent 3,137,588, 1964.
81. Hart, R. T., U. S. Patent 3,311,492, 1967.
82. Hart, R. T., U. S. Patent 3,332,794, 1967.
83. Hart, R. T., U. S. Patent 3,425,896, 1969.
84. Stewart, D. T. and McAlpine, I., British Patent 1,337,983, 1973.
85. Stewart, D. T. and McAlpine, I., British Patent 3,741,782, 1973.
86. McAlpine, I., "Ammonium zirconium Carbonate, An alternative Insolubilizer for Coating Binders," *TAPPI 1982 Coating Conference Proceedings,* TAPPI PRESS, Atlanta, 1982, pp. 165–179.
87. "AZC insolubilizer for starch/latex paper coating formulations," Data Sheet 103B, Magnesium Elektron, Flemington, NJ, 1984.
88. Extensive literature is available from Magnesium Elektron, Flemington, NJ, Particularly recommended are Data Sheets 103A-H.
89. Sharif, S. "Zirconium Compounds as Starch and Protein Insolubilizers for the Paper Manufacturing Industry," Zirconium Technology Corp., Midland, TX, 1990.
90. Blumenthal, W. B., *The Chemical Behavior of Zirconium*, D. Van Nostrand, Princeton, NJ, 1958, p. 38–39.
91. Ibid., p. 344.
92. "Zirconium-polymer interactions in the paper and board industry," Data Sheet 103G, Magnesium Elektron, Flemington, NJ, 1984.
93. "Coating structure and water retention improvements with AZC," Data Sheet 103F, Magnesium Elektron, Flemington, NJ, 1984.
94. "Zirconium chemicals in the paper industry — an introduction," Data Sheet 103A, Magnesium Elektron, Flemington, NJ, 1984.
95. "AZC insolubilizer for protein/latex paper coating formulations," Data Sheet 103C, Magnesium Elektron, Flemington, NJ, 1984.
96. "AZC insolubilizers for high latex paper coating formulations," Data Sheet 103B, Magnesium Elektron, Flemington, NJ, 1984.
97. Von Tegge, G., *Staerke* 12: 321(1960).
98. Bochert, P. J., *Kunstoffe* 51: 137(1961).
99. Mehltretter, C. L., in "Production and Use of Dialdehyde Starch" (Whistler, R. L. and Paschall, E. F., Eds.), *Starch: Chemistry and Technology,* Academic, New York, 1985, Chapter XVIII.
100. Mehltretter, C. L. and Wise, C. S., *Ind. Eng. Chem.* 51: 511(1959).
101. Mehltretter, C. L., U. S. Patent 2,830,941, 1958; Chem. Abstr., 1958, 52, 15312.
102. Pfeifer, V. F., *et al., Ind. Eng. Chem.* 52: 201(1960).
103. Mehltretter, C. L., *Staerke* 15: 313(1963).
104. Mehltretter, C. L., *Methods Carbohydr. Chem.* 4: 316(1964).

105. Guthrie, R. D., *Adv. Carbohydr. Chem.* 16: 106(1961).

106. Radley, J. A., *Starch Production Technology,* Applied Science, London, 1976, pp. 423–448.

107. Mehltretter, C. L., *Staerke* 18: 208(1966).

108. Rutenberg, M. W. and Solarek, D., in *Starch Chemistry and Technology,* 2nd edn. (Whistler, R. L., BeMiller, J. N., and Paschall, E. F., Eds.), Academic, Orlando, 1984, p. 366.

109. Weakley, F. B., *et al., Tappi* 44(7): 456–459(1961).

110. Ernst, A. J., *et al., Tappi* 45(8): 646–650(1962).

111. Weakley, F. B. and Mehltretter, C. L., U. S. Patent 3,188,223, 1965.

112. Carr, M. E., *et al., Tappi* 49(6): 244–248(1966).

113. Zienty, M. F., U. S. Patent 3,324,057, 1967.

114. Suzumura, H. and Miyahara, H., U. S. Patent 3,324,057, 1967.

115. Curtis, J. H., U. S. Patent 3,395,106, 1968.

116. Weakley, F. B., *et al., Staerke* 24(6): 191–194<1972).

117. Verbanac, F., U. S. Patent 4,060,506, 1977.

118. Young, A. H. and Verbanac, F., U. S. Patent 4,079,025, 1978.

119. "Starpol 100 Polymerizable Starch," A. E. Staley Mfg. Co., Decatur, IL.

120. Young, A. H., Verbanac, F., and Protzman, T. F., *Ultraviolet (UV) Cure of Modified Starch Coatings and Adhesive Binders,* Chem. Prop. Crosslinked Polym., Proc. ACS Symp., 1977, pp. 191–203.

121. Bramel, G. F., A. E. Staley Mfg. Co., Decatur, IL, personal communication, 1988.

122. Tessler, M. M., U. S. Patent 4,153,585, 1979.

123. Zimmerman, J. L., U. S. Patent 3,884,853, 1975.

124. Hinterwaldner, R., *Coating* 19(10): 364–368(1986); Chem. Abstr., 1990, Vol. 106, No. 12, Sec. 142, Abstr. No. 086273.

125. Paschall, E. F., Thayer, F. D. Jr., and Minkema, W. H., U. S. Patent 3, 132,113, 1964.

126. Thayer, F. D. Jr., U. S. Patent 3,127,393, 1964.

127. Cescato, R. W., U. S. Patent 3,654,263, 1972.

128. Taylor, K. G., U. S. Patent 3,719,514, 1972.

129. Reference 108, p. 358.

130. Gramera, R. E. and LeRoy, D. H., U. S. Patent 3,4649,974, 1969; Chem. Abstr. 1970, 72,22777.

131. Ritson, D. D. and Jen, Y., U. S. Patent 2,918,438, 1959.

132. Taraka, M. and Nonaka, M., Japanese Patent 7107607, 1971; Chem. Abstr. 1972, 76, 61181a.

133. Strother, G. W. and Coker, W. P., U. S. Patent 3,607,331, 1971.

134. Kawakami, S., *et al.,* Japanese Patent 84129229, 1984; Chem. Abstr., 1985, 101 (22), Sec. 143, Abstr, No. 194023.

135. (a) Kawakami, S., *et al.,* Japanese Patent 85055028, (1985); (b) Kawakami, S., *et al.,* Euro Pat. EP81994, 1983, CA 99 (10) 72452r.

136. Nakanishi, M., *et al.,* European Patent Application 0062338, 1982.

137. Suzuki, K., Tanaka, H., and Senju, R. *J. Japan Wood Res. Soc.* 21(8): 473–479(1975); Abstr. Bull. Inst. Pap. Chem. 1976, 46 (11), p. 1176.

138. Kress, B. H., U. S. Patent 2,968,581, 1961.

139. Kress, B. H., U. S. Patent 2,868,773, 1959.

140. Perkins, G. H., U. S. Patent 3,125,455, 1964.

141. McKnight, G. S. and Brown, R. C., U. S. Patent 2,865,773, 1958.

142. Trouvé, Claude, "Glyoxal et Impression Offset (Glyoxal and Offset Printing)," Proceedings of the 37th Conference of the Technical Association of the Paper Industry, Paris, FR., Nov. 6-9, 1984.

143. *Pocket Pal® — A Graphic Arts Production Handbook,* 12th edn. (Brown, M. H., Ed.), International Paper Co., New York, 1979.

144. *The Printing Ink Manual,* 3rd edn. (Bisset, D. E., Goodacre, C., Idle, H. A. Leach, R. H., and Williams, C. H., Eds.), Northwood, London, 1979.
145. Graham, P. M. and Kriski, T. L., U. S. Patent 4,421,564, 1983.
146. Personal communication from Alfred Lamy, Director of Marketing, Ultra Additives, Paterson, N.J., May 10, 1990.
147. "Zinplex 15 Material Safety Data Sheet," Ultra Additives, Paterson, N.J., 1990.
148. "Zinplex 15 Technical Data Sheet," Ultra Additives, Paterson, N.J., 1990.
149. Belche, J. R. Jr. and Ellis, G. C., *Tappi* 40(2): 94–99(1957).
150. Ritson, D. D. and Landes, G. C., *Tappi* 38(5): 300–303(1955).
151. Cotton, F. A. and Wilkinson, G., *Advanced Inorganic Chemistry,* 4th edn., John Wiley & Sons, New York, 1980, pp. 591–600.
152. Miller, G. D., Air Products and Chemicals, Allentown, PA, personal communication, 1990.
153. James, A. B., Sequa Chemicals, Technical Service Dept., Chester, S.C., personal communication, 1990.
154. Shulman, J. J., "Behind the Contents," *FLEXO* 15(4): 6(1990).
155. "Industry News," *FLEXO* 15(4): 88(1990).
156. "Innovative Perspectives on Coating and Sizepress Technology," Data Sheet, Sequa Chemicals, Chester, SC, 1990.
157. TAPPI Technical Information Sheets, 1993.
158. Blokhuis, G., "Testing the Surface Strength of Paper," IGT Publication 35, 1979.
159. TAPPI T 459 om-88, "Surface strength of paper (wax pick test)."
160. TAPPI Useful Method 507, "Pick resistance test."
161. TAPPI Useful Method 534, "Coating pick tests using fiber bond tester."
162. TAPPI Useful Method 487, "Ink rub test of paper (Sutherland Tester)."
163. Berilla, J. J., "Dry-Rub Testing Techniques," TAPPI 1972 Coating Additive Committee Meeting Notes, TAPPI PRESS, Atlanta.
164. Lee, H. H., "The crockmeter — a means of measuring wet-rub resistance," *Tappi* 57(3): 113–115(1974).
165. Jansen, F. B., "Applications of the IGT – Printability Testers," *IGT Monograph 12,* 7th revised edn., 1980, pp. 29–30.
166. TAPPI T 458 om-84, "Surface wettability of paper (angle of contact method)."
167. Unpublished Test Procedure PC84-004, Sequa Chemicals, Inc., Chester, S.C.
168. Proof-Bau Bulletin, Proof-Bau, Dr.-Ing. H. Durener gmbh, AlCH 23 D-8123, Speissen, Germany.
169. Jansen, F.B., "Application of the IGT-Printability Testers," IGT Monograph 12, seventh revised edition, 1980, pp. 34-35.
170. Eckhart, C.G., "Advantages of Co-Binder Coatings for Varied Printing Processes," *Tappi* 58(9); 73-6(1975).
171. TAPPI Useful Method 513, "Wet Pick Test with Proofpress and Blade Attachment."
172. TAPPI Useful Method 468, "Wet Rub of Coated Paper (Wet Finger)" Formerly RC 185.
173. Montague Machine Co. Bulletin, "Adams Wet-Rub Tester."
174. Adams, F. R., "A Method to Determine the Degree of Wet Rub Resistance," *Tappi* 43(11): 923-7(1960).
175. TAPPI Useful Method 462, "Wet Rub of Coted Paper (Brush in Taber Abrader)." Formerly RC 184.
176. Black, W.C., "A Method for Comparing the Wet-Rub Resistence of Coated Papers," *Tappi* 32(10):467-8(1949).
177. Floyd, W.C. and Sharif, S., U.S. Patent 5,268,030, 1993.
178. "Sequarez 82, Zirconium-Based Binder Insolubilizer" Data Sheet, 1994, Sequa Chemicals, Inc., Chester, S.C.

CHAPTER 9: PRESERVATIVES *(1)*

Stephen D. Goldstein and Robert W. Cade

Introduction

Coated paper production in the United States has increased from 5,860,000 short tons in 1985 to 8,563,000 short tons in 1994 *(2)*. This trend toward an increased demand is expected to continue, driven by market demand for the specialized sheet characteristics imparted by the coating process. The sheet characteristics which are affected by coatings include improvement in the brightness, opacity, and printability of the finished product. With a rising demand for coated paper and paperboard, an anticipated increase in the number of specialized paper grades, and the trend toward alkaline mill conversions the need for quality coating will increase. However, the chemical makeup and physical properties of many coatings form an environment conducive for microbial growth. Microbial spoilage of coating components and formulation is becoming a more economically significant problem for the paper industry. Reasons for this include the increased cost of coating ingredients, increased downtime losses associated with today's larger and faster coating operations, and the greater emphasis on high quality standards in coated paper and paperboard.

Microorganisms can have a negative impact on coating operations in two ways — coating application and spoilage. Application may be adversely affected by microorganisms plugging flow meters, fouling nozzles, and forming deposits which can slough off coating pan surfaces onto the paper sheet. Spoilage problems may result in changes in viscosity, pH, color, rheology, and the development of malodors or lumps in the coating formulation or their ingredients. These serious problems can be minimized, or possibly avoided, by proper microbiological control procedures, including the use of chemical preservatives as coating additives.

Uncontrolled growth of bacteria and fungi normally present in water, binders, pigments and other coating ingredients is responsible for spoilage of the coatings. The presence of these organisms and the adverse effects of their rapid growth have frequently been demonstrated. Spore-forming bacteria in dry-end coatings can adversely impact the production of food grade paperboard products required to meet the Dairyman's Standard *(3)*. As a rule, it is neither economically feasible nor necessary to eliminate these organisms by sterilization of the coating components. Instead, spoilage problems can be prevented by procedures that keep the microbiological population within acceptable limits. Such preservation methods include the use of good housekeeping and effective coating preservatives as well as optimized storage practices.

Coating preservatives can be divided into two principal types: those which actually kill the organisms (biocides) and those which inhibit growth without actually killing (biostats). Biostats may be advantageous under conditions where minimal contamination exists. However, as specialty grades proliferate, longer storage times and more demanding conditions for specific coating components may increase the need for biocides over biostats.

Categories of Coatings

Coating formulations generally contain one or more components able to support microbial growth under appropriate conditions. The types of coatings and the number of components contained in a coating formula, over time, have become more numerous and more complex. Historically, the manner and extent of coating spoilage could be tied to the type of binder utilized in the coating. There are three basic types of binders: starches, proteins, and synthetics. This section is principally concerned with the effects of microbial degradation on the binder components. However, other coating additives and components may be subject to microbial attack. These components will be discussed later in this section.

Starch Binders

The most commonly used binder and coating formulation is starch. Starch is a carbohydrate that is highly susceptible to degradation by a variety of microorganisms. Starches are composed of dextrose polymers or polysaccharides, mainly amylose and amylopectin. Natural starches generally contain a much higher percentage of amylopectin, as compared to amylose. Microorganisms enzymatically degrade these complex molecules into simple sugars. The sugars are then further degraded, resulting in the production of carbon dioxide and acids. The effects of microbial action on coatings containing starch include pH depression, viscosity changes, retrogradation, and off-odors.

Dry, unmodified starch is only minimally susceptible to microbial growth and its detrimental effects. It is the introduction of makedown water that starts the process of biological growth and degradation either by acting as a source of contamination or by lowering solids enough for existing organisms to thrive and/or multiply. All starches and starch-containing coatings are subject to spoilage to some degree. Enzymatically converted starches are more easily attacked by microorganisms, as the starch is already partially degraded. The heating process that is used to prepare starch slurries in coatings is usually sufficient to kill many organisms present in these solutions. However, bacterial endospores will normally survive and, after the coatings have cooled, readily germinate, resulting in vegetative cells capable of degrading the starch. Quench water used to dilute and/or cool cooked starch can reinoculate cooked starch and promote endospores to germinate.

One immediate result of starch spoilage in coatings is pH depression. The acids produced by the microbial activity can drive the retrogradation of the starch to the amylose form. The presence of excess amylose in a coating affects the viscosity and quality and, therefore, the uniformity of sheet surface application *(4)*.

Protein Binders

Protein binders are frequently used in combination with other binders, such as starch. The proteins most used as binders are casein and soy protein, although other proteins can also be utilized. As with starch, proteins are polymeric chemical substances composed of amino acid sub-units instead of sugar molecules.

Microorganisms attack the protein molecules, cleaving them into amino acids that can be readily absorbed by the organism. An immediate result of this degradation process, as with starch degradation, is the formation of simple organic acids that can precipitate the remaining intact protein. The presence of anaerobic bacterial action in a protein or starch coating may result in the production of putrefactive odors.

The effects of microbial attack on protein-containing coatings includes changes in pH, viscosity, and coating stability. A small increase in acidity can result in a very large reduction in viscosity of the coating. Usually, the principal result of pH depression is the precipitation of protein, which destroys the uniformity and stability of the coating.

Synthetic Binders

Synthetic binders are often used in combinations with starch or protein binders. However, there are coating formulations that use only synthetic latex or combination of synthetic binders. Synthetic binders include styrene-butadiene, vinyl acetates, and acrylates.

Coating formulations containing only synthetic binders may still be susceptible to microbiological spoilage. Depending upon the chemical structure involved, it is possible for organisms to hydrolytically break down polymeric structures. It is also possible that microorganisms can attack and degrade the dispersing and stabilizing components in latex formulations. Synthetic binders are generally rendered most susceptible to spoilage as a result of dilution (or makedown) with biologically contaminated water or coating components.

Latex-containing coatings that have spoiled are usually more difficult to stabilize in terms of viscosity and uniformity, and may have a tendency to foam. However, pH and viscosity swings are not as common or as severe as with coatings containing natural components. Frequently, spoilage of latex coatings is not easily detectable. Direct observation of poor coating performance is usually the first, and best, indicator that biological spoilage of the coating may have occurred.

Other Components

Coating components other than binders, such as dispersants and defoamers, may also be subject to microbial attack. Nearly all additives currently used in coating formulations are biodegradable to some extent, meaning they are subject to breakdown to simple components utilizable by microorganisms as nutrient sources. For example, some coating additives contain fatty acids and fatty acid derivatives that are readily metabolized by microbes.

The mechanism by which organisms degrade a particular compound will depend upon its structure and the types of organism(s) involved. Also, depending upon the component involved, the results of microbial spoilage may vary. In general, the result of microbial degradation of any coating or component is the reduction or the elimination of its intended effect *(5)*.

Spoilage Microorganisms

The microbial spoilage of coating, coating components, and other chemical additives commonly used in papermaking processes is caused by a variety of bacteria and fungi. Typically, no one species is found to be responsible for the spoilage, but rather a mixed population of bacteria and/or fungi exists in a spoiled product. For example, a spoiled batch of coating might contain several species of aerobic bacteria, as well as several species of yeast and/or filamentous fungi. This diversity is due to the ability of many microorganisms to degrade and use the organic and inorganic compounds (e.g., starches, proteins, and acetates) found in coatings and coating formulations.

Types of Spoilage Microorganisms

Bacteria

Aerobic bacteria. Bacteria which, in respiring, can use O_2 as their final electron acceptor and thus require an environment containing free oxygen. Some examples of aerobic bacteria include: *Pseudomonas* spp. and *Acetobacter* spp.

Anaerobic bacteria. Bacteria which in respiring use organic or inorganic compounds (other than O_2) as their final electron acceptors. Some of these bacteria, such as *Desulfovibrio* spp., can only metabolize compounds in the absence of free oxygen and are classified as "strict anaerobes." Others, including *Klebsiella, Enterobacter, Serratia,* and *Proteus* spp., can grow in the presence or absence of free oxygen and are classified as "facultative anaerobes."

Spore formers. Bacteria, which when stressed by environmental changes (e.g., high temperatures, toxic chemicals, and dehydration) form resistant endospores. The endospores are able to survive environmental stress and later germinate to a vegetative state. *Bacillus* spp. and *Clostridium* spp. are examples of bacteria capable of forming endospores.

Mold

These include species of genera *Penicillium, Aspergillus, Trichoderma, Fusarium, Chaetomium, Endomyces,* and *Mucor.*

Yeasts (and Yeast-like Fungi)

These include species of genera *Saccharomyces, Candida, Rhodotorula, Cryptococcus,* and *Monilia.*

Microbial spoilage of coatings and coating components can be caused by a variety of circumstances. For example, bacterial contamination in coatings can frequently be traced to the makedown water. The organisms present in the water multiply rapidly upon entering the nutrient-rich coating, resulting in a spoiled product. Anaerobic conditions often result in pipelines and storage tanks in which coatings/ additives are allowed to remain without agitation. This generally occurs as a result of aerophilic/aerotolerant organisms utilizing the available O_2, reducing the dissolved oxygen levels to where anaerobic metabolism can proceed. Fungal contamination often results from the use of contaminated coating ingredients, i.e., starches, proteins, etc. Poor cleaning of process equipment and storage tanks is another factor that can contribute to the microbial spoilage of coatings.

In order to prevent microbial spoilage, it is often necessary to locate the source of the contamination. Once the source has been identified (via microbiological tests), it can be treated accordingly and the problem of microbial spoilage thus significantly reduced or eliminated.

Spoilage Mechanisms

The following is a general description of how microorganisms degrade natural compounds (starches and proteins) and synthetic compounds (latex). Because of the diverse metabolic pathways used by microorganisms, the descriptions have necessarily been generalized for this discussion.

Starch

Starch, the most common component of paper coatings, can typically make up 40% of the total solids in a coating. Microorganisms readily degrade starch, which is composed of polysaccharides (22–26% amylose and 74–78% amylopectin). In order to degrade the polysaccharides into simple sugars which the microorganisms can metabolize, the microorganisms must secrete enzymes. These microbially produced enzymes hydrolyze the starch into relatively small molecular weight sugars utilizable by the microorganism. The resulting microbial starch degradation and subsequent metabolism of the sugars may produce various acids, gases and/or malodorous by-products. These by-products often produce quite drastic changes in the viscosity, odor, and pH of a starch-containing coating. A contaminated starch-containing coating might experience a pH decrease of 2 units or more in less than a 24-hour period.

Some starches, for example derivatized and chlorinated starches, have chemical groups different from those of natural starches and these may be more resistant to microbial degradation. Conversely, dextrinized, enzyme-converted or chemically converted starches are already partially degraded and may be more prone to microbial attack. Often unmodified starch, due to economic considerations, is purchased by the mill and later modified (converted) using either enzymes or chemicals such as ammonium persulfate, sodium hypochlorite or hydrogen peroxide. Typically, the starch is heated following modification (enzyme or chemical) and this normally eliminates most microorganisms. However, heat-resistant bacterial endospores can survive and cause spoilage after they germinate when the temperature is reduced.

Viscosity and pH are two physical properties of starch-containing coatings that can be affected by microbial spoilage. Microbial enzymatic depolymerization and/or acid production can greatly reduce starch solution viscosity. However, microbial spoilage of some coatings can result in viscosity increases. Usually this occurs when microbial acid production interferes with the dispersants found in the coating component slurries. In summary, viscosity can increase or decrease in a coating as a result of microbial spoilage, depending on what coating component is impacted.

Protein Spoilage

Proteins, such as casein and soy protein, are often used in coatings either as binders or pigment dispersants. When used as binders, the proteins are often used in combination with other latices. Regardless of their function in coatings, proteins can be easily degraded by microorganisms. The proteins are ultimately broken down to amino acids which are used by the microorganism to synthesize new proteins. The microbial degradation of proteins can result in various by-products including acids, gases, and putrefactive compounds. Anaerobic bacteria are particularly adept at breaking down proteins and producing putrefactive by-products believed to be responsible for the off odor of the end product.

As in starch degradation, protein degradation via microorganisms results in changes in the physical properties of the protein-containing coating. The physical changes can include pH changes and viscosity changes. Most indicative of protein degradation (putrefaction) in a coating are the malodors and color changes produced via microbial by-products. As the acidity of the coating increases, proteins will precipitate and cause a change in the coating color.

Synthetic Spoilage

Synthetic compounds, i.e., synthetic latices and organic compounds, are now found in many coatings. These compounds are frequently used in conjunction with proteins and starches to form the binder portion of the coating. However, there are coatings which contain binders composed entirely of a synthetic compound. Typical synthetic binder systems include polyvinyl acetate, polyvinyl alcohol, polyacrylates, various resins, and styrene butadiene. These synthetic/semi-synthetic binders are subject to microbial attack just like their starch and protein counterparts.

Microbial degradation of synthetic binders alone in coatings is not as obvious as degradation involving starch or protein binders. Microorganisms are capable of degrading either the polymer itself or the emulsion stabilizing portion of the latex system. The binders are more prone to microbial attack following their dilution or addition to coating formulations.

Of the synthetic compounds used as binders, latex is one of the most common, and its susceptibility to microbial spoilage needs to be specifically addressed. Unlike the spoilage of coatings containing starch/protein binders, the spoilage of latex binders does not usually result in as significant a change in coating pH or viscosity. Also, there is usually little or no color change or malodor present. Some indicators of a spoiled latex coating are: significant foaming, less stability to shear forces, and, in more severe cases, difficulty in maintaining viscosity. Quite often a spoiled latex coating goes unnoticed until testing of the sheet shows evidence of a coating problem. A spoiled latex coating may produce poor coating adhesion to the base sheet, low pigment holding capacities, or undesirable supercalendering properties.

Spoilage of Other Coating Components

The starch, protein, and synthetic components of adhesive systems are not the only components of coatings which can be microbially degraded. Flow modifiers, defoamers, lubricants, dispersants, colloids, and other coating components/additives can also be microbially degraded and thus affect the coating's performance. Even though these additives usually represent only a small percentage of the overall coating, their microbial degradation results in the destruction of the additive's function. For example, a polyphosphate used as a dispersant in a coating could be enzymatically converted to an orthophosphate via the metabolic activity of microorganisms. The resulting orthophosphate could not maintain the pigment dispersion necessary and the resulting coating would be low in pigment. Usually, the dispersants, i.e., polyphosphates, are added in quantities large enough so that microbial enzymatic conversion is not a factor. However, if the coating is stored for prolonged periods, at temperatures conducive to microbial growth, microbial spoilage may result.

Some additives, such as defoamers, emulsifiers, and flow modifiers, contain fatty acids and fatty acid derivatives. These are excellent nutrient sources which are readily degraded by microorganisms. Oil-based defoamers are particularly prone to microbial attack and subsequent enzymatic breakdown of their hydrocarbon components. When these additives are degraded by microorganisms, the result is that the additive is unable to perform its desired function. Thus the desired foam control, flow characteristics, or emulsion stability may be reduced or lost entirely. Many of the newer defoamers, additives, etc. contain synthetic compounds in place of these oils, and this reduces their susceptibility to microbial attack.

Factors Affecting Microbiological Spoilage

All living organisms, including the microorganisms which cause coating spoilage, require certain essential growth factors. These growth factors, provided by the organism's environment, include water, nutrients, and a hospitable setting. Understanding the environmental factors that can contribute to microbiological growth is the first step in setting up good spoilage prevention and control procedures.

Water

A principal necessity of all forms of life is water. Microorganisms can extract water from nearly any source imaginable, including their food and the atmosphere. Coating formulations, as well as the individual components of the coatings, will always contain sufficient water to support biological growth. If the water content of a coated paper sheet approaches 15% or more, or if the coated paper is stored in a humid environment, there may be sufficient water present to support the growth of microorganisms.

Nutrients

As noted earlier in this chapter, nearly all components currently used in coating formulations are subject to biodegradation. The list of potential food sources is sizable: starch, protein, latexes, defoamers, lubricants, resins, emulsifiers, and others. In fact, most coating formulations represent a virtual cornucopia of nutrients for microbiological growth.

Temperature

The optimum temperature range for active growth by most microorganisms is 10–50°C, or 52–122°F. Coating formulations are frequently subjected to temperatures exceeding the upper limit of this range. Heating formulations to temperatures exceeding 60°C contributes significantly to control biological growth and, thus, spoilage. However, there are organisms (thermophiles) that can thrive at these temperatures and bring about coating spoilage. In addition, some bacteria are capable of forming endospores, which are highly resistant to both heating and desiccation. These spores survive temperature extremes until conditions become favorable for germination and return to active growth.

pH

The optimum pH range for bacterial growth is between 6 and 8, with an outside range of 4 to 9. Fungi can also grow over a wide range of pH, but optimum growth occurs between pH 3 and 7. In an acidic environment, fungi tend to grow actively while bacterial growth will generally tend to be naturally inhibited. The difference in pH optima for various organisms can be significant, e.g., in stock systems. However, pH conditions of most coatings are generally neutral to alkaline, which facilitates the growth of nearly all types of organisms and the possibility of rapid microbiological spoilage. The primary concern in coating systems is the possible negative effect of pH on preservative efficacy. Alkaline pH can bring about the hydrolysis and dissociation of many coating preservatives. As a result, some preservatives will require extremely high concentrations to effectively inhibit biological growth in neutral to alkaline coating systems. Some biocidal preservatives may actually require concentrations so excessive that they are no longer cost-effective, cause odor problems, or tend to be incompatible with the coating.

Atmosphere

Microorganisms are classified with reference to respiratory metabolism into three primary categories: aerobic, anaerobic, and facultative. Aerobes require environmental oxygen as a terminal electron acceptor in their energy-generating system. Anaerobic organisms are generally poisoned by oxygen and therefore utilize other ions as the terminal electron acceptor in their energy generation. Facultative organisms are able to metabolize and grow in the presence or absence of oxygen, growing more vigorously in the presence of oxygen.

Aerobic organisms will grow at any location in a coating system that has substantial aeration resulting in a high level of dissolved oxygen. Growth of aerobic and facultative organisms can drop the dissolved oxygen level to allow the growth of obligate anaerobes. Anaerobic organisms will be found at any point in a system that has poor agitation, permitting stagnation to take place. Anaerobes can also be found underneath deposits where these organisms are shielded from an oxygenated environment. Facultative organisms will grow throughout a coating system.

Incubation

The rate or time required for a coating to spoil will depend on a combination of all the factors discussed previously in this section. In general, the longer a quantity of coating is permitted to remain in a condition suitable for biological growth, the more likely the coating could be lost to spoilage. Control measures, including maintaining temperatures over 145°F and the use of a preservative, tend to retard biological growth in a coating. Still, while control measures may slow down spoilage of coating formulations, if quantities are stored for periods of 4–8 hours or more, spoilage is still possible *(6)*.

Preventing Coating Spoilage

Protecting a system from the influx and growth of microorganisms will in all cases be the most effective and cost-efficient means to avoid problems resulting from the end products of microbial metabolism. Since this is not completely achievable, various control programs exist. These programs are especially important with regard to coating formulations since no practical means to restore degraded coating have yet been developed.

As discussed, many coating components are excellent food sources for many microorganisms. Even many of the inorganic additives such as clays, titanium dioxide, and carbonates which by themselves will not readily support microbial life contain components (i.e., phosphate type dispersants) which stimulate growth.

Effective preservative programs which minimize the impact of microbial contamination will generally incorporate more than one control method. The most successful of these programs often use a combination of good housekeeping practices in conjunction with process water sanitation and preservative additions.

Five major factors will be considered here to understand coating spoilage prevention:

1. Microbiological monitoring
2. Process water and additive cleanliness
3. Routine system boilouts
4. Controlling operating parameters
5. Preservative selection/addition.

Microbiological Monitoring

The method used to culture various types of microbes is important since not all problem species may be cultured by the same method. Routine monitoring of the Total Bacteria Count (TBC) on one type of medium, incubated at one standard temperature, may not provide an accurate assessment of bacterial number and species diversity. For example, anaerobic (oxygen-sensitive) bacteria which produce acidic wastes and may be capable of producing CO_2, H_2 gas, and/or H_2S will not grow under standard conditions used for general microbiological monitoring. The various requirements of different species which may contaminate coating formulations must be taken into account when monitoring the microbiological status of the formulation.

Two indicators used to quickly evaluate coating quality are viscosity and pH. The most common problems are manifested by a reduction in pH to less than 8.7, which can result in amylose retrogradation. This can result in a drastic reduction of viscosity.

Some bacteria can produce acidic by-products which lower the pH of coatings. By regularly monitoring the formulation pH, steps can be taken to correct the problem before it results in a severe viscosity change or spoilage. Changes in pH are easily monitored and will provide an early warning of potential spoilage before microbial populations reach high levels.

Process Water and Additive Cleanliness

Controlling the major entry points of microorganisms into the formulation is the first line of defense in an effective preservation program. The single most important source of additive contamination comes from contaminated makedown water. If the fresh water supply is contaminated and not sufficiently treated, one can eventually expect to develop coating deterioration problems. Fortunately this source of inoculum is also the easiest and most cost-effective component of the coating formula to control. Maintaining an effective concentration of free Cl_2 residual will eliminate this area of concern.

As previously mentioned, coating components themselves are often a source of microorganisms. Regardless of preservative addition, prior to shipping it is still possible for formula components to contribute active microorganisms.

Each type of preservative is effective over a specified period of time, given the parameters under which it must work. The pH value of the component can influence the preservative's effectiveness over time. Hydrolysis of many of the organic compounds used as preservatives will occur more rapidly in an alkaline environment resulting in products with less biocidal efficacy. This is also the same pH environment which will be conducive to bacterial growth.

Since storage is often at ambient temperatures for extended periods of time, this too will accelerate growth.

Routine System Boilout

Once a coating spoils, it must be discarded. To avoid spoilage (odor, pH and viscosity changes) of the next batch, the system must first be cleaned before initiating a preservative program. The best method to avoid microbial re-contamination is to boil out the system. Otherwise, the new coating formulas will be contaminated by the previous coating which had already developed a significantly high microbial population.

Boilouts should be scheduled on a routine basis as part of a preventative program. An acid or alkaline cleaning solution should be recirculated throughout the entire system from the raw materials storage tanks to the coating equipment so that all microbiological deposits are removed. This eliminates the possibility of "reseeding" the system and negating the benefits possible from a thorough boilout.

Once the solution has circulated to allow removal of the deposits, the system is drained and flushed with fresh water. An initial flush with fresh water containing hypochlorite is an easy and low-cost method to ensure a microbiologically acceptable system on startup.

Controlling Operating Parameters

Although controlling operating parameters (e.g., equipment design) and environmental factors (e.g., temperature and pH) will not eliminate coating spoilage, it does provide a means to minimize the impact of microbial contamination. Preventing long storage times is especially important.

Smaller holding tanks provide the ability to have more frequent coating turnover. The reduced dwell time in the tank prevents the microorganisms from reaching high numbers which could negatively impact the coating quality before it could be used.

Dual or multiple small tank systems provide the option of alternating tanks so that older, contaminated coating left in the tank bottom can be washed out before refilling. If a tank becomes contaminated and the microbial populations increase during storage, any coating remaining serves to contaminate the new batch at higher initial levels, thus speeding up the fouling process.

Equipment should also be protected from airborne contamination. Tanks should be accessible but covered to prevent dust and debris from entering and yet allow convenient washup. Air ducts should also be redirected away from exposed coating at the machine to reduce contamination when recirculated to a use tank.

Preservative Selection/Addition

Spoilage control is not always possible by only using the previously described methods. Often an effective chemical preservation program is necessary. Biocidal compounds, when properly selected and applied, are capable of inhibiting bacterial and fungal growth in the papermaking system. Many of these compounds are often used in the preservation of coating formulations.

Selecting the appropriate chemical control depends on product compatibility, degree of required preservation, types of microbes to be controlled, physical design of the system, storage time, and environmental conditions of temperature and pH.

Choice of a preservative involves evaluating the inhibiting properties of the compounds against the types of microbes presently causing spoilage problems. As practiced in monitoring methods, it is important to compare the ability of the preservatives to control problem microbes including thermophilic and anaerobic bacteria. The latter frequently impart odors from the production of volatile fatty acids (VFAs).

The VFAs most often screened are: acetic (C2), proprionic (C3), butyric (C4), and valeric (C5). These VFA levels can increase in the coating loop from coater to use tank.

The interaction between any system additive and a preservative should also be considered when selecting a chemical control method. Since physical and chemical compatibilities must be compared in the process, it is a simple procedure to "bench test" the compounds for any color or viscosity change, precipitation, loss of stability, or objectionable odor.

Once the preservatives are screened for compatibility and inhibition properties, it must be applied so that it effectively contacts the problem organisms. A uniform blending of preservative and coating therefore is critical. Application can be made either neat or diluted, depending on the ability of the preservative to incorporate into the formula or its components.

Many biocide actives used in the papermaking process are also appropriate for coating applications and have approval by the U.S. Food and Drug Administration (FDA).

However, for those coating formulations which require food grade sanction, the list of usable preservatives is more limited. These chemical compounds are listed in the Code of Federal Regulations, Title 21, Subchapter B, Sections 176.170, 176.180, and 176.230. This list has been fairly stable over the years, with few changes due to stringent governmental procedures.

Preservative Use

In cases when it is not possible to obtain adequate control of spoilage by the use of good housekeeping procedures and fresh water chlorination, treatment with an effective preservative is necessary. A variety of chemical compounds are available that can inhibit and control the growth of bacteria and fungi in coating materials. No single compound has been universally accepted or found to be entirely satisfactory for all coatings or operating conditions.

Chemical preservatives currently in use include organo-sulfur compounds, organo-bromine compounds, heterocyclic nitrogen compounds, and other organic or inorganic substances. Some common organic biocidal actives used to preserve coating compounds are methylene bis thiocyanate; beta-bromo-beta- nitrostyrene; isothiazoline; 2-bromo, 2-nitropropane 1,3 diol and 3,5-dimethyl-1,3,5,2H-tetrahydro thiazine-2-thione.

The selection of the appropriate preservative and the methods of application are dependent on, among other things, the severity of the spoilage problem, the nature of the materials to be preserved, and the degree of preservation required. Other factors that need to be considered are the types of microorganisms involved, methods of mixing or dispersing, and environmental conditions such as temperature and pH. The chemical and physical compatibility of the preservative with other coating ingredients, storage times, coating specifications, and susceptibility of coating to spoilage also need to be addressed. The treated coatings should be monitored for the following: pH, temperature, viscosity, color, odor, and emulsion stability.

One important criterion in the selection and preservation of coatings used on paperboard for food contact use is the status of the chemical compound under the various U.S. Food and Drug Administration (FDA) regulations. These are listed in the Code of Federal Regulations, Title 21, Subchapter B, Sections 176.170, 176.180, and 176.230 (2). The number of FDA-allowed coating preservatives is not expected to increase greatly in the near future because clearance procedures have become increasingly stringent and complex. The most commonly used biocide actives for food grade coating preservation are:

- 3,5-dimethyl-1,3,5,2H-tetrahydrothiazine-2-thione
- glutaraldehyde
- 5-chloro-2-methyl-4-isothiazolin-3-one and 2-methyl-4-isothiazolin-3-one.

Microbiological Evaluation of Preservatives

At present, there is no official TAPPI Test Method for evaluating preservatives, however, there are two methods, ASTM Designation E823-85 (bacterial) and ASTM Designation E875-88 (fungal), which are designed to evaluate preservatives for aqueous-based products used in the paper industry *(7)*.

The following is a test method, based upon the ASTM methods, which can be used to determine the concentration of a preservative required to inhibit the growth of microorganisms (fungi and/or bacteria) in a coating or any other substrate.

As mentioned previously, the physical/chemical compatibility of the preservative with the coating/substrate should be ascertained prior to starting the microbiological evaluation. The addition of preservatives which are incompatible with the coating/substrate can result in changes in the physical properties of the coating/substrate which could greatly impair its function.

Apparatus Required

1. Balance — sensitive to 0.1 g at a load of 200 g. An analytical balance sensitive to 0.1 mg should be employed to weigh the candidate preservative.
2. Bottles — borosilicate glass milk dilution bottles with either screw caps or Escher rubber stoppers.
3. Incubators — capable of maintaining temperatures of 28—37°C; 50°C required for some thermophiles. Anaerobic incubators are necessary if it is desirable to evaluate the preservative against anaerobes.
4. Petri Dishes — 100 [MULT. SIGN] 15 mm, plastic or borosilicate glass.
5. pH Measurement — any reliable pH meter is suitable to standardize the pH of the culture medium. Nonbleeding test strips are recommended for samples.
6. Pipets, 1.1 ml milk dilution type, 1.0 ml graduated in 0.01 ml, and 10 ml graduated in 0.1 ml.
7. Sterilizers — Steam sterilizer and hot-air oven.

Microbiocides and Materials

1. Freshly prepared test solutions (usually 1.0% or 0.1%) of the preservative.
2. Test Materials/Substrate — Freshly prepared pigment slurries, adhesives, dye, rosin, polymer, sizing solution, and/or other materials should be used as the substrate.
3. Culture medium.
   - Tryptone Glucose Extract Agar for bacteria.
   - Sabouraud (with dextrose) Agar, Mycophil, or Potato Dextrose Agar for fungi.
   - It may be necessary, since some test materials do not support rapid microbial growth, to add supplemental nutrients to the test material. However, these supplemental nutrients may increase the level of biocide needed or possibly inactivate a normally active biocide. Some test materials may not support fungal growth; these materials most likely do not need a preservative to prevent fungal growth.

Test Organisms

1. The test inoculum will vary depending upon the material to be preserved. The specific spoiled material will serve as the inoculum. The viability of the microorganisms existing in the spoiled material should be verified by making spread cultures on the agar medium used in the test procedure.
2. To provide a uniformly inoculated substrate, the spoiled inoculum should be added to the entire quantity of the test substrate at one time, mixed thoroughly, and then divided into separate test aliquots.
3. The material under test should be inoculated with sufficient spoiled material to give a bacterial count of > 1,000,000/ml or higher. For fungi, enough spoiled material should be added to the test material to give a measurable amount of growth on a spread plate.

Procedure

1. Dispense 50 g aliquots (or any suitable quantity) of the inoculated test material aseptically into sterile milk dilution bottles or other suitable container. Set up controls in duplicate.

2. Make the following additions aseptically to each bottle in the order named and shake vigorously after each addition, using 20 complete cycles in a vertical motion.
   - Add the desired volume of the stock solution of the antimicrobial to be tested to give the desired concentration in parts per million or percent. Do not add biocide to the controls. Include in each test a minimum of 5 concentrations of the antimicrobial under test. Typical preservative concentrations range from 50 to 300 ppm, however, concentrations may vary, depending on the material tested.
   - Stock solutions of the antimicrobial should be of such strength so that the volume of the antimicrobial solution added is no more than 1% of the total volume of sample in each bottle. Record the pH of all samples at the beginning of the experiment, using nonbleeding test strips, and other physical characteristics such as significant change in color, odor, viscosity, etc.
3. **For fungi only:** Prepare spread cultures of the test material before and after inoculation to determine the sterility of the original test material and level of fungal contamination after inoculation. Incubate plates at 28–30°C for 96 h or until the control plates show sufficient growth for rating.
   - Level of growth can either be a growth/no-growth rating, or some type of rating scale such as 0 to 4 with 0 being no growth and 4 being heavy growth.
   - Incubate all samples at 28°C (or other temperature related to the optimal growth temperature) with the bottles loosely capped. At weekly or other suitable time intervals, depending on the rate of growth of the fungi in the test material, mix each sample and immediately spread on Sabouraud's Agar and incubate at 28°C for 96 h. Observe each sample bottle for the evidence of spoilage such as visual growth of fungi, changes in color, odor, pH, etc. Record the growth of the fungi for the streaked samples.
   - Rechallenge all treated samples after streaking with 2 ml of inoculum obtained from one of the duplicate controls.
4. **For bacteria only:** Incubate all samples at 32°C (or suitable temperature) with the milk dilution bottle loosely capped.
   - Make serial dilution plate counts initially before and after inoculation for the controls and weekly for all samples to determine the number of survivors. Reinoculate all treated samples with 2 ml from one of the duplicate control samples as the rechallenge. (Approximately 2 ml is the estimated weekly loss due to evaporation.)
   - Each preservation experiment is based upon the use time.
5. Each preservation experiment, whether fungal or bacterial, is based upon the use time of the material to be preserved. Continue the test for at least the average length of time that the specific type of product would need to be protected (generally 6 weeks) unless the material under test is preserved for a shorter period of time, for example, one week. Record the nature and extent of spoilage based upon the criteria used.
6. After each incubation period, mix each sample thoroughly and, immediately: 1) plate in Tryptone Glucose Extract Agar (for bacteria) according to standard dilution plating techniques, and incubate at 32°C for 48 h or 2 days) spread on Sabouraud Dextrose, Mycophil or Potato Dextrose Agar (for fungi) using standard spread plate methods and incubate at 28°C for 96 h. The first method is prepared because it is more reliable.

A spread plate may not recover contamination at low levels (e.g., spores present that one does not see) to sample and streak.

7. At the completion of the incubation period: (a) count the number of bacterial colonies at the dilution that yields a count between 30 and 300 colonies, and record the number of bacterial colonies per ml, or (b) record the relative growth of fungi on the streak plates.

Calculation (for the bacterial test only)

At each sampling time and at the end of each test, calculate the number of bacteria inhibited or killed at each microbiocide concentration tested as follows:

$$\%\ inhibition = \frac{(control\ plate\ count - test\ plate\ count)}{control\ plate\ count}\ x\ 100$$

A 95% or more decrease in count is indicative of the effectiveness of the antimicrobial under test. Visual deterioration and other signs of degradation, such as changes in pH, color, viscosity, etc., should also be used to judge the degree of preservation.

Results

At each sampling time and at the end of each test, record all results. For the bacterial test, an effective preservative is one which maintains a 95% kill/inhibition throughout the duration of the test. Also there should be no physical evidence of spoilage in the test substrate. For the fungal test, an effective preservative is one which reduces the fungal growth and maintains that reduction based upon results from the streaked plates. The study is not valid unless fungal spoilage is evident in the control based on visual fungal growth, obvious changes in physical characteristics or other reliable criteria.

Non-plate Count Methods for Monitoring Coatings for Microbial Growth

In recent years, many test methods have been developed which are capable of rapidly estimating the number of microorganisms present in a system/deposit. At present there is no official method using a rapid count procedure for measuring microbial populations in coatings/papermaking additives. However, since these rapid count microbiological test procedures are frequently used to analyze other substrates found in papermaking processes, they should be mentioned accordingly.

Generally, rapid count procedures measure microbial populations in systems/samples via the use of oxidation/reduction indicators (redox dyes). For example, a rapid count method might use a redox dye such as resazurin or TTC (triphenyltetrazolium chloride). As the redox dye is microbiologically reduced, a color change occurs and the time required for this change is proportional to the number of metabolizing (viable) bacteria present. So, based upon the time necessary to reduce the dye, one can estimate the number of microorganisms present.

Some rapid count methods use other reagents and measure microbial populations via other mechanisms. For example, one such method measures total ATP (adenosine triphosphate) which is used to estimate the number of microorganisms in the sample.

The rapid count methods — as do most methods — have their specific limitations. For example, samples containing high suspended solids or fiber are difficult to measure using rapid count tests unless they are filtered. Also certain microorganisms such as fungi and filamentous bacteria do not react as rapidly as unicellular bacteria do with the redox indicators and therefore are difficult to detect.

Regardless of their limitations, these rapid count methods offer alternatives to plate count methods for evaluating preservatives in coatings/additives. However, at present, the only generally accepted methods are those using plate count procedures.

Mold-Resistant Paper and Paperboard

The growth of fungus (mold) on paper or paperboard is quite common, especially under conditions of high moisture. Fungal presence can originate from contamination (via air, soil and/or water) of the paper/paperboard after it is manufactured. Also, it is quite possible that the finished paper/paperboard may be contaminated (i.e., via spores or from the coating).

Grades of paper/paperboard which are consumed quickly or are not exposed to long-term contamination and incubation are not generally degraded by fungal organisms. But the following grades often require special treatment to impart resistance to fungal and sometimes bacterial growth:

- Acoustical tile
- Decorative wallpaper
- Flooring felt
- Food and soapboard
- Hardboard
- High moisture gypsum liner
- Industrial tissue
- Latex specialty
- Molded pulp
- Secondary packaging
- Secondary specialty
- Specialty bag
- Specialty wrapping
- Wet label.

In order to make paper/paperboard resistant to fungal growth, a chemical preservative (biocide) must be added to the product. The presence of preservative in the finished product will provide resistance to varying degrees, depending on the quality, efficacy, and persistence of the anti- microbial compound(s). Of course, the more favorable the conditions for contamination and fungal growth, the more challenged the preservative is to inhibit such growth.

Certain fungi are pigmented and may cause discoloration of the paper or pulp. Often these pigments will pass through coatings and latexes (i.e., of flooring tile) and show up as spots of various colors causing definite quality problems.

Mold inhibiting agents are generally applied from 1 to 20 lb/ton of paper or board produced, depending upon objectives and grades being produced. Application is generally made to the stock or applied via size press, spray bar, roller or calendar stack. Coatings may also need preservation in some cases.

Predictive tests and performance tests of mold inhibition agents are usually necessary to ensure effective treatment. Both TAPPI and ASTM methods have been used. Sometimes tests individually designed by a paper mill or a user may be specified for the inhibited product. The mold inhibition test most commonly referenced at this time follows: TAPPI T 487 pm-85 "Fungus resistance of paper and paperboard."

Types of agar used are as follows: mineral salts, mycophil, sabouraud dextrose, and potato dextrose.

Some of the organisms that are used in mold inhibition testing follow: *Aspergillus niger, Aspergillus terreus, Aspergillus oryzae, Penicillium expansum, Penicillium funiculosum, Chaetomium globosum,* and *Trichoderma viride.*

Performance tests and guarantees involved in mold inhibition vary widely from mill to mill and grade to grade. For example, some flooring tile companies provide a five-year guarantee against mold growth for their products. Some grades may be mold-proofed with little justification and testing.

Literature Cited

1. Lutey, R. W. and May, O. W., "Preservatives," *Paper coating additives,* TAPPI PRESS, 1978, Chap. 7, pp. 106–123.
2. *Pulp & Paper Fact Book,* 1995 North American edn., Miller Freeman Publications, pp 178-185.
3. Bendt, H. T., "How to control bacteria in food grade paperboard production," *Pulp & Paper,* November 1985.
4. Poock, S. J., "Microbial Contamination in the Use of Starch", *Tappi J.*, August 1985.
5. Smook, G. A., *Handbook for Pulp and Paper Technologists,* Joint Textbook Committee, Atlanta and Montreal, 1982.
6. Block, S. S., *Disinfection, Sterilization and Preservation*, Lea & Febiger, 1983.
7. ASTM 1989, Vol. 11.04.

Acknowledgment

The authors extensively referenced the chapter prepared by the previous authors, Richard W. Lutey and Oscar W. May. The authors thank those who reviewed this chapter, including T. G. Surface, product manager—biocides, Pulp & Paper Group, Nalco Chemical Company and Dr. C. L. Wiatr, Ph.D., microbiologist and group leader, Pulp & Paper Group, Nalco Chemical Company.

Subject Index